Air Pollution in the United Kingdom

Air Pollution in the United Kingdom

Edited by

G. Davison and C.N. Hewitt
Lancaster University, UK

THE ROYAL
SOCIETY OF
CHEMISTRY
Information
Services

Sep/ae
Cher

The proceedings of a joint one day meeting on 'Air Pollution in the United Kingdom' held on 23 September 1996 at University of Lancaster

Special Publication No. 210

ISBN 0–85404–767–0

A catalogue record for this book is available from the British Library

Published by the Royal Society of Chemistry,
Thomas Graham House, Science Park, Milton Road,
Cambridge CB4 4WF, UK

Typeset by Computape (Pickering) Limited, North Yorkshire
Printed by Bookcraft (Bath) Ltd

SD 11/24/97 *futher*

Preface

Man has been aware of the fact that his industrial, agricultural and social activities have adverse effects on the chemical composition of the atmosphere, with consequent adverse environmental and health effects, for centuries. However, it is only comparatively recently that sufficient effort and resources have been devoted to the study of air pollution to allow quantitative assessments of these effects to be made.

The United Kingdom is unique in being an island state with a long industrial and post-industrial economic history, with a high population density, with a varied climate and with an extensive array of legislation and legal precedent referring to air pollution in place. In addition, the country has experienced a wide range of air pollution problems, from the poisonous sulfurous winter smogs of the 19th and early-mid 20th centuries, to the deposition of 'acid rain' in upland rural areas, to the wintertime episodes of high nitrogen dioxide concentrations still seen in our cities, largely due to motor vehicles, to the late 20th century phenomenon of rural photochemical ozone pollution. Industrial, agricultural, societal and legislative changes have all had effects on the type and amount of air pollution suffered in the country. All these factors in totality make the United Kingdom an ideal 'case study' for assessing air pollution.

Very recently, fully automated, state-of-the-art, monitoring networks have been established across the United Kingdom providing data of the highest possible quality for a range of air pollutants. Although relatively modest in scale, and distinctly lacking in spatial coverage in places, these networks now deliver sufficient data to allow a reasonable understanding of the causes, concentrations, and to a lesser extent, the effects, of air pollution to be made in the country.

This book results from a symposium held in September 1996 at Lancaster University organised jointly by the Environmental Chemistry Group and the NW Region of the Analytical Division of the Royal Society of Chemistry on 'Air Pollution in the UK'. It aims to give a brief, but authoritative, overview of air pollution in the country, drawing in part on the data provided by the national monitoring networks. It is not totally comprehensive, omitting, for example, full consideration of heavy metals, persistent organic pollutants or radioactive air pollutants. Nevertheless we hope it will be of value to all those interested in assessing, preserving and enhancing the environmental quality of this beautiful land.

Nick Hewitt and Gerry Davison
Lancaster

Contents

Indoor Air Pollution

D.R. Crump

Building Research Establishment, Watford, Herts WD2 7JR UK

1 Introduction

Indoor air pollution is the presence within buildings of toxic or other substances which may, directly or indirectly, be a cause of ill health or discomfort. The term is usually used in the context of non-industrial buildings where there are not normally any processes that result in exposure to substances at concentrations approaching the occupational exposure limits referred to by health and safety regulations, such as the Control of Substances Hazardous to Health Regulations (COSHH) in the UK. Exceptionally such limits are exceeded and the occurrence of about 100 deaths per year in the UK due to carbon monoxide release from faulty gas appliances is of particular concern[1]. Other exceedances could occur during works such as painting and other decorating under poorly ventilated conditions[2,3]. However the more common situation in non-industrial buildings is for some pollutants to occur at concentrations higher than in the outdoor air as the result of emissions from products in the building, but at concentrations far below maximum exposure limits.

This paper outlines the main types and sources of indoor air pollutants and their effects on health. It discusses the importance of exposure of people to air pollutants in non-industrial indoor environments as a contributor to the total exposure to air pollutants. Current knowledge about levels of indoor air pollution in the UK is discussed and the results of a study of benzene in indoor air presented because of the particular concerns for health effects due to exposure to this compound.

2 Types and Sources of Indoor Air Pollutants

Table 1 summarises the main types of indoor air pollutants and their sources. Volatile organic compounds (VOCs) encompass a wide range of compounds and various studies of indoor air quality have reported more than 250 compounds at a level exceeding 1 ppb[4]. Particulate air pollution is also complex, both physically and chemically. The main determinant of the

Table 1   *Sources and types of indoor air pollution*

Source	Main Pollutants
Combustion of fuel	carbon monoxide (CO), nitrogen oxides
Tobacco smoke	VOCs, particulates, CO
People	carbon dioxide (CO_2), organic compounds (body odours)
Building materials	VOCs, formaldehyde, radon, fibres, particulates
Consumer products	VOCs, formaldehyde, pesticides
Furnishings	VOCs, formaldehyde
Office equipment	VOCs, ozone (O_3)
Outdoor air	sulfur dioxide (SO_2), nitrogen dioxide (NO_2), O_3, particulates, viable particulates, benzene
Contaminated land	methane, VOCs
Washing and cleaning, people, outdoor air	water (producing degradation and mould growth if condensation)
Animals	allergens

behaviour of an atmospheric particle is its size and classifications of types are made on the basis of aerodynamic diameter[5]. Particulate matter indoors has also been classed according to its origin into 5 classes; plant (pollen, spores), animal (bacteria, viruses, hair, skin cells, insect parts and products), mineral (asbestos, carbon, clays, elemental particles and man-made mineral fibres), combustion (tobacco smoke, cooking, heating appliances and industrial plants) and radioactive (radon decay) products which attach to larger particles[6]. A wide range of volatile and semi-volatile organic compounds have been found to be absorbed to particulate matter indoors including formaldehyde, n-alkanes, fatty acids, phthalate esters and polyaromatic hydrocarbons[7,8,9].

Sources of indoor pollutants can be classed as those due to materials present in the buildings, those due to people and their activities and those due to pollutants generated outdoors. The consequence of there being indoor and outdoor sources is that some pollutants occur at higher concentrations indoors because they are emitted within the building, some have a similar concentration indoors and out because they enter the building via infiltration and ventilation, and others are at a lower concentration indoors because during entry and within the building they are deposited on or adsorb to surfaces and are removed from the indoor air. For most buildings there are typical ranges for the ratio of concentrations of individual pollutants in indoor and outdoor air, variations within the range being due to specific factors such as source strengths, building characteristics, occupant behaviour and level of outdoor pollution.

Table 2 shows typical indoor-outdoor (I/O) concentration ratios for some common indoor pollutants in buildings in developed countries[10,11]. In the absence of specific indoor sources, concentrations of carbon monoxide, nitrogen dioxide, sulfur dioxide, ozone and particulate matter are equal to or lower than concentrations outdoors. For the wide range of organic compounds emitted by many products in buildings the indoor-outdoor ratio is considerably higher than one.

Table 2 *Indoor-outdoor concentration ratios for some common indoor air pollutants*

Pollutant	Typical I/O Ratio
CO_2	1-3
CO; no indoor sources	1
indoor sources	1-5
NO_2; no indoor sources	0.5-1
indoor sources	2-5
SO_2	0.1-0.6
O_3	0.1-0.8
Particulate;	
Total suspended	<1
Inhalable (< 2.5 µm), no smoking	1
Total volatile organic compound concentrations (TVOCs)	7
Decane	19
Undecane	20
Benzene	3
Toluene	6
Xylenes	6
1,2,4-trimethylbenzene	15
Limonene	80

3 Personal Exposure

People spend a large part of their time in indoor environments and the greater part of this is in the home. Certain groups of the population such as infants, the elderly and sick may be at home for more than 90% of their time. Hence exposure to indoor air pollutants can account for a major part of the total exposure of many people to air pollutants. This has been demonstrated to be the case for a number of different pollutants in studies involving personal monitoring as well as stationary monitoring of pollutants in different micro-environments in which people spend time. Table 3 summarises the main findings which show personal exposure concentrations to be similar to those in the home and often greater than outdoors.

For some pollutants, concentrations associated with transport can be higher than indoors, but the time spent in a vehicle for most people is much less than time in the home. For the carbon monoxide study it was commented that exposure during transport was probably significant. The available studies are not comprehensive for pollutant types, countries or population groups and further work is required. The study of particulates compared night time and day time exposures showing higher personal exposures in day time and this was thought to be the result of activities resulting in a 'personal cloud' of particulates in the breathing zone. The studies reporting TVOC concentrations used different definitions of TVOCs and the values are not directly comparable across studies but the within-study differences between indoor, personal exposure and outdoor air show the importance of indoor air for personal

Table 3   *Studies of personal exposure to pollutants and concentrations in different microenvironments*

Pollutant	Concentration $\mu g\ m^{-3}$			
	Personal	*Indoor (home)*	*Outdoor*	*Transport*
NO_2 [12] (homes with gas cooking, Manchester UK)	32.7	38.5[a], 60.7[b]	20.5	–
NO_2 [12] (homes with electric cooking, Manchester, UK)	26.7	24.6[a], 28.1[b]	26.6	–
CO [13] (Washington, USA)	6.7	–	3.2	–
CO [13] (Devon, USA)	10.0	–	6.6	–
Particulates (PM_{10})[14] (daytime, USA)	150	95[a]	95	–
Particulates (PM_{10})[14] (night time USA)	77	63[a]	86	–
Formaldehyde[15] (Australia)	21	23.7[a], 23.5[c]	2.4	–
Benzene [16] (USA)	17.6	15.8[a], 16.7[b]	7.7	–
TVOC [17] (USA)	2900	1000	500	–
TVOC[18] (UK)	280	170[a], 220[b]	18	350
Benzene[18] (UK)	6	6[a], 6[b]	3	20
Toluene[18] (UK)	17	10[a], 15[b]	6	85
Undecane[18] (UK)	8	7[a], 8[b]	1	5

a = living room; b = main bedroom; c = kitchen

exposure. A further study of seven people undergoing 25 common activities such as washing clothes, using deodorisers, painting and smoking showed that activities increased exposure to VOCs often by factors of 10, sometimes by factors of 100, compared to exposures during sleeping[19].

4 Health Effects

Table 4 summarises the health effects that could be caused by indoor air pollutants and gives some examples of particular types of pollutant that could produce those effects. This is drawn from an extensive published review which discusses the current knowledge of health effects and the uncertainties involved with assessing the significance of exposure to the relatively low levels of pollutants present in indoor air compared with the industrial occupational environment[20]. Combustion products, environmental tobacco smoke (ETS) and biological contaminants are the main agents associated with respiratory health effects indoors. Allergic asthma and extrinsic allergic alveolitis are the two most serious immunological diseases caused by allergens in indoor air. The principal agents present in indoor air associated with lung cancer are ETS and radon decay products. Other known or possible human carcinogens found in indoor air include asbestos, polycyclic aromatic hydrocarbons (PAHs), benzene, formaldehyde, some pesticides and nitrosamines. Irritation of tissues of the skin and mucosal membranes can cause inflammation resulting in a sensation of heat (calor), redness (rubor), swelling (tumour), pain (dolor) and a

Table 4 *Possible health effects of indoor air pollutants*

Type	Example	Pollutants
Respiratory effects	pulmonary function change respiratory symptoms sensitisation of airways spread of respiratory infections	ETS (NO_2), viral infection, legionella
Allergy	allergic asthma entrinsic allergic alveolitis allergic rhinoconjunctivitis (humidifier fever)	house dust mites, cockroaches (skin scales, faecal particles), pets (hair, saliva, urine, skin), pollen, moulds
Carcinogenic and reproductive effects	lung cancer leukaemia mesothelioma mutagenicity human reproduction	ETS, radon, chemical carcinogens (benzene), asbestos
Irritative effects	inflammation of skin and mucous membranes	formaldehyde, VOCs, ETS, NO_2, O_3, fibres
Sensory and other effects on the nervous system	odour dry skin neurotoxicity	VOCs, formaldehyde, pesticides
Cardiovascular and other systemic effects	cardiovascular disease hepatic effects	CO, ETS
Multiple chemical sensitivity	mild discomfort total disability depression	chemicals

certain loss of function (functio laesa). Principal agents are formaldehyde and other aldehydes, VOCs, ETS and the temperature and humidity which may influence the level of irritation. Sensory effects are perceptual responses to environmental exposures resulting in a conscious experience such as odour, touch, itching and freshness. Different sensory perceptions are combined into perceived comfort and into the sensation of air quality. Exposures to ETS and carbon monoxide have been implicated in cardiovascular symptoms, and in changes in cardio-vascular disease morbidity and mortality. Multiple chemical sensitivity is a chronic multisystem disorder and affected persons are frequently intolerant to some foods and react adversely to some chemicals at levels generally tolerated by the majority of people.

Guidance on acceptable levels of indoor air pollutants for the avoidance of health effects is available for a limited number of substances. The World Health Organisation published recommended air quality guidelines for indoor and outdoor air for 16 organic compounds and 15 inorganic substances which have been widely quoted and are currently under review to take account of research since 1987[21,22]. For compounds reportedly without carcinogenic effects, the guidelines are based on the lowest observed adverse effect level and for malodorous substances separate guideline values are given for sensory effects. For carcinogenic compounds no guideline value is given but the risk associated with lifetime exposure to 1 μg m^{-3} of the agent is estimated. There are also a number of national guidelines such as those in Canada for 8 substances plus water vapour[23], in Japan, for carbon monoxide and airborne particulate matter[24] and Norway for VOCs (400 μg m^{-3}), formaldehyde (100 μg m^{-3}), carbon monoxide, nitrogen dioxide, particulates, fibres and mites[25]. Many countries including the UK have specific guidelines for radon[26].

Guidelines have also been proposed for mixtures of organic compounds in terms of TVOCs to avoid sensory discomfort to building occupants[27]. The National Health and Medical Research Council in Australia recommended a value of 500 μg m^{-3} with no single compound contributing more than 50% of the total[28]. The Finnish Society of Indoor Air Quality and Climate have suggested target values for concentrations of TVOCs, formaldehyde, ammonia, carbon monoxide, ozone, radon, total suspended particles, carbon dioxide and odour intensity[29].

5 Indoor Air Pollution in UK Homes

5.1 Introduction

The Department of the Environment (DOE) has responsibility within the UK Government for indoor air quality in non-occupational environments. It sponsors research on indoor air quality, to assess the risks to public health and well being, to inform the development of policy and to provide advice to the general public and building owners. As part of this work, DOE commissioned the Building Research Establishment (BRE) to undertake a study of levels of specific pollutants in a sample of normally occupied homes and this was called the BRE Indoor Environment Study. This was carried out in collaboration with the University of Bristol (Institute of Child Health) who were carrying out a longitudinal survey to investigate problems of child health and development called the Avon Longitudinal Study of Pregnancy and Childhood (ALSPAC). It has involved the monitoring of specific indoor air pollutants in 174 homes in the county of Avon over a 12 month period. The main aims of the work were to provide data on the range of concentrations of these pollutants occurring in UK homes and to identify the factors such as household characteristics and occupant activities, which influence the level of pollutants.

Once the monitoring study in Avon was complete, the DOE commissioned the Medical Research Council's Institute for Environment and Health (IEH) to perform comprehensive reviews of the exposure and health effects of the pollutants measured. These reviews were based on other published literature as well as the BRE study and were presented at a workshop of international experts. The reviews and recommendations from the workshop have been published[30].

The BRE and IEH studies are part of the DOE strategy for improving indoor air quality in UK homes which has three stages; 1) determine baseline levels of indoor air pollutants in homes, 2) the interpretation of results of monitoring to assess the implications for health and well being and 3) the development of a range of suitable remediation techniques. To date the strategy to improve indoor air quality has focused on the 6 main pollutants measured in the BRE study. While further research is required on all of these, the recommendations made by the IEH are in the process of being implemented[31]. The range of pollutants under consideration is being increased with the aim of expanding the strategy to include carbon monoxide, PM_{10} and ETS.

This section summarises a BRE report of the Indoor Environment Study[32] and presents results for benzene in some detail. Benzene is a human carcinogen and there is a UK air quality standard for ambient air[33].

5.2 The ALSPAC-BRE Indoor Environment Study Structure

Women were recruited to the ALSPAC study in early pregnancy with the intention of monitoring the health of the children for seven years from birth. The county of Avon was selected for study because it has a mixture of urban and rural areas, varied industry and a population which is demographically representative of the rest of Britain. To be eligible for the study, subjects had to be residents in the study area while pregnant and have an expected date of delivery between 1 April 1991 and 31 December 1992; 14,893 mothers were recruited to the main ALSPAC study.

The BRE Indoor Environment Study involved 174 households of participants of the ALSPAC study. Approximately 10 women were recruited each month from November 1990 to March 1992 from a list of volunteers supplied to BRE by the University of Bristol. Participants completed initial and postnatal questionnaires, and up to 12 monthly updates about their homes and activities. The initial questionnaires were completed in conjunction with the placing of various samplers to measure concentrations of pollutants inside and immediately outside the home. The following pollutants and potential allergens were measured; formaldehyde, TVOCs and selected individual compounds, nitrogen dioxide, airborne bacteria, airborne fungi and house dust mites in furnishings.

5.3 Methods of Measurements

Samples of formaldehyde, VOCs and nitrogen dioxide were collected by passive samplers. For the first sampling event in each home the samplers were set out by a researcher and their operation explained to the householder. Thereafter they were sent to the household and returned to the laboratory by post. Samples of bacteria, fungi and house dust mites were collected by active techniques requiring visits to homes by researchers. Details of the measurements are as follows:

5.3.1 Nitrogen Dioxide. Nitrogen dioxide was measured using Palmes diffusion tubes with a 14-day exposure period. Samplers were placed in the kitchen, main bedroom, living room and outside, four times a year at 3-month intervals.

5.3.2 Mites. Mite numbers were counted in living room and bedroom carpets in 35 homes every 2 months for a year. A specially adapted vacuum cleaner head was used and a 1 m^2 area vacuumed. Mites were extracted from the dust by flotation using kerosene and ethanol/water mixtures.

5.3.3 Fungi and Bacteria. Samples of airborne fungi and bacteria were collected by membrane filtration in the living rooms of 163 homes four times a year. The material collected on filters was removed and spread onto agar plates and incubated. In addition some airborne samples were taken using a six stage Andersen sampler.

5.3.4 Formaldehyde and VOCs. Formaldehyde was measured using GMD 570 Series Dosimeters which were placed in the main bedroom and lounge of all 174 homes and in other rooms and outdoors in a sub-sample. When returned to the laboratory the reagent (2,4-dinitrophenylhydrazine) was desorbed from the sampler and formaldehyde was determined using High Performance Liquid Chromatography (HPLC). Samplers were set out for a 3-day period each month over a 12-month period. VOCs were measured using a stainless steel adsorbent tube packed with Tenax TA. The tubes were placed in the vicinity of the formaldehyde dosimeters and exposed for 28 days. When returned to the laboratory they were analysed by thermal desorption and gas chromatography. Samplers were replaced each month with the objective of undertaking 12 consecutive sampling periods in each home.

5.4 Summary of Published Results

5.4.1 Nitrogen dioxide. Mean kitchen concentrations of nitrogen dioxide were higher than those in living rooms and main bedrooms. Two homes had at least one 14-day mean above the 150 µg m^{-3} WHO guideline value[21] for 24 hours. Levels in the home were significantly affected by outdoor concentrations. Autumn and winter months and more urban environments were associated with highest concentrations. Mean concentrations outdoors in winter were 25 µg m^{-3} compared with 17.5 µg m^{-3} in summer. Where there was no indoor source, indoor levels tended to be lower than outdoors. Gas cooking was the dominant source of indoor nitrogen dioxide, heating and tobacco smoking having effects which are probably negligible by comparison. Natural gas used as the main cooking fuel was, in some locations and seasons, a more significant factor than outdoor concentration.

Mean concentrations in the kitchen of homes with a gas cooker were 4–8 µg m^{-3} higher than the outdoors, depending on the season. In homes with electricity cookers mean concentrations in kitchens were 1.6 to 10.5 µg m^{-3} lower than the outdoors, indicating a loss of nitrogen dioxide indoors through adsorption to surfaces.

5.4.2 Mites, Fungi and Bacteria.

The geometric mean mite count was 186 mites m^{-2} carpet in living rooms and 175 mites m^{-2} carpet in the bedroom. The maximum number of mites was 16,800 mites m^{-2}. Over 60% of counts were above 100 mites m^{-2} carpet. *Dermatophagoides pteronyssinus* accounted for 95% of the mites recovered. Mite numbers showed a seasonal fluctuation with a general increase in numbers in summer months with the peak occurring in September.

The geometric mean fungal count was 235 cfu m^{-3} air and for bacteria was 366 cfu m^{-3} air. *Penicillium* was the most frequently isolated and abundant fungi. *Bacillus* was the predominant bacteria in the 163 homes, but *Staphylococcus* and *Micrococcus* dominated the air samples collected in the 35 homes using the Andersen sampler. Fungal counts were highest in summer months as were bacteria to a lesser extent.

5.4.3 Formaldehyde and Other VOCs.

Indoor concentrations exceeded those outdoors with formaldehyde being 12 times higher and TVOCs ten times higher which shows that there are important indoor sources of these compounds. There was no significant difference between the concentrations of pollutants measured in the living rooms and main bedrooms.

There was a strong relationship between the mean formaldehyde concentration in homes and the age of dwelling. The mean formaldehyde concentrations increased the newer the dwelling and the mean concentration in homes built since 1982 was about three times higher than those built before 1919. Of the 12 homes where a single 3-day mean reading exceeded 100 µg m^{-3}, which is equivalent to a 30-minute World Health Organisation (WHO) air quality guideline value[21], eight were built since 1982. There was a seasonal fluctuation in concentration; formaldehyde concentrations in summer months were higher than those in winter.

A statistically significant relationship was found between the TVOC concentration indoors and the occurrence of painting activity during the period of sampling. The presence of a smoker who smoked more than 50 cigarettes per week was associated with significantly higher TVOC concentrations compared with homes with no smokers. Typically 85 individual compounds were detected by the analysis of the adsorbent tube and the summation of these makes up the TVOC value. A summary of the concentration data for three compounds, benzene, toluene and undecane, was included in the BRE report to illustrate the relative importance of indoor and outdoor sources of these compounds. The present paper now presents detailed results for benzene.

5.5 Results of Measurements of Benzene

Table 5 summarises the results of measurements of benzene in the 174 homes and the 14 outdoor sites. Mean indoor concentrations in different rooms were between 1 and 1.6 times values at 13 outdoor sites. In 8 homes (or 5% of those studied) the mean benzene concentration in at least one room exceeded the UK air quality standard[33] of 16 μg m^{-3} (as a running annual average). In 48 homes at least one monthly mean reading was above 16 μg m^{-3}. The mean indoor concentration of 8 μg m^{-3} in homes in Avon is similar to the mean of 10 μg m^{-3} for homes reported by a World Health Organisation working group who produced a frequency distribution for benzene concentrations and other VOCs based on field studies in Germany, Italy, the Netherlands and the United States[34]. The concentrations measured outdoors are within the range reported for urban concentrations at 9 sites in England during 1994/95 measured with continuous monitors and similar to the six monthly mean concentration (4 μg m^{-3}) measured using diffusive samplers at one site in Bristol for the period July to December 1992[35].

Table 5   *Summary of benzene measurements (μg m^{-3}) in homes in Avon*

Room	Mean of monthly readings in each house					
	n	mean	SD	RSD (%)	minimum	maximum
Main bedroom	173	8	4	58%	2	32
Living rooms	173	8	6	59%	2	46
Kitchens	6	6	2	35%	3	8
Bathrooms	6	6	3	44%	3	8
Second bedrooms	20	5	2	49%	2	11
Outside	13	5	1	29%	3	8

RSD = relative standard deviation; SD = standard deviation

As a result of the large amount of information gained from participants completing questionnaires it was possible to analyse the data for associations between benzene concentrations and various characteristics of the house, its location and occupant behaviour. The full range of information available and details of statistical tests applied were the same as those applied to the data for TVOCs and formaldehyde[32]. Table 6 summarises the results of tests using Student's t-test for the difference of two means and one way analysis of variance (ANOVA) for comparison of more than two means which showed a statistically significant difference between means with a 95% confidence level.

Benzene concentrations in the main bedrooms are influenced by whether the home is located in a city or rural location. This is the consequence of higher concentrations of benzene in the outdoor air in the city than in rural areas as shown in Table 7. The concentration at rural sites is 53% of that in the city and this is due to the influence of motor vehicle emissions as reported by other workers[33]. The same trend is present for living rooms, but is not statistically

Table 6 *Benzene – summary of statistically significant (95% confidence) relationships between benzene concentrations and household characteristics and occupant activities*

Factor	Benzene Concentration $\mu g\ m^{-3}$				Statistical Test and Results
	Main Bedrooms		Living Rooms		
	mean	SD	mean	SD	
1. Location					t - test
rural	6.40	4.18	(8.53	8.91)	city > rural - main
city	9.03	3.09	(9.21	3.36)	bedrooms only
2. Garage attached	9.73	6.52	10.02	7.15	t - test
garage attached					attached garage >
no garage or garage not attached	6.87	3.30	7.47	4.96	no attached garage
3. Use of car with attached garage car in attached garage	12.02	7.60	12.30	8.48	t - test car in garage > car not in garage
car never in attached garage	6.58	2.36	6.90	2.75	
4. Smoker living in home	(8.18	4.26)	9.58	7.32	t - test
smoker	(7.06	4.37)	7.15	4.23	smoker > no smoking -
no smokers					living rooms only
5. Heavy smoker (50+ cigarettes/week living in home heavy smoker	9.18	4.93	9.99	6.47	t - test heavy smoking >
no smokers (or smoking visitors)	7.13	4.53	7.28	4.41	no smoking
6. Months home unoccupied for 3 or more days home occupied	7.78	4.17	8.07	5.12	t - test home occupied >
unoccupied for > 3 days	5.85	3.80	6.44	4.78	not occupied

( ) difference between means not statistically significant

significant. The difference between living rooms and main bedrooms may be due to smoking of tobacco in the home which acts as a confounder. The presence of a smoker results in a significant increase in benzene concentrations in the living room but has less effect on main bedrooms. Possibly fewer cigarettes are smoked in the main bedroom but the study did not collect information on this. If more than 50 cigarettes per week are smoked by an occupant then benzene levels in living rooms and main bedrooms are significantly higher than in homes where there are no smokers. Mean concentrations in no-smoking homes are 73% of that in homes with heavy smokers. Similar results have been reported by other studies of benzene concentrations in homes

Table 7   *Mean concentrations of benzene at outdoor sites in Avon*

Location Type	Number of sites	Mean concentrations $\mu g\, m^{-3}$	SD
rural	3	3.6	1.0
suburban	3	5.0	1.2
urban	4	5.0	0.9
city	2	6.8	1.2

in West Germany and the United States which found median concentrations in homes without smokers were 6.5 and 7.0 $\mu g\, m^{-3}$, respectively, compared with concentrations of 11 and 10.5 $\mu g\, m^{-3}$ in homes with smokers[36,37].

A further factor associated with higher concentrations of benzene in living rooms and main bedrooms is the presence of an attached garage. Homes without an attached garage have a mean concentration that is 70–75% of that in homes with an attached garage. This is probably due to the evaporation of fuel from the motor vehicle and also the running of the engine in close proximity to the home. The importance of the car as a source of benzene is demonstrated by comparing homes with attached garages that do not keep the car in the garage and homes with attached garages that are used for the car; the presence of the car results in mean concentrations being 80% higher. A study of sources and concentrations of benzene in seven homes in the United States found that at homes with a garage or environmental tobacco smoke, mean indoor and personal exposure concentrations were 2 to 5 times higher than outdoor levels at all but one home[16]. Benzene levels in the 4 attached garages included in the study ranged from 3 to 196 $\mu g\, m^{-3}$ and usually were higher than indoor living areas or personal exposure levels.

Comparing mean concentrations of benzene in homes during months that the home is occupied with months where the occupants were away for a period of 3 or more days shows the influence of occupant activities on benzene concentrations. Concentrations during months when occupants were away for some period are 75% of that when homes were occupied nearly all of the time. This effect must be due to a combination of factors, notably the absence of smoking and a car in an attached garage. It is also possible that benzene is released as the result of other activities although there was no significant effect found for activities such as painting, new furnishings and use of some other products. However numbers of cases to assess the influence of these factors were small. Some evidence for other possible sources is provided by the highest monthly readings reported; house 49 in June (106 $\mu g\, m^{-3}$) and house 97 in October-January (45–81 $\mu g\, m^{-3}$). These high readings coincide with high TVOC values (1.14 mg m^{-3} in 49 and 2.03 mg m^{-3} in 97). In house 49 toluene was the dominant VOC and there was a group of very volatile compounds that could not be identified. House 97 gave a very different chromatogram with significant peaks due to C_{10} and C_{11} aliphatic hydrocarbons and a very large peak due to p-dichlorobenzene. Toluene was also present as the fourth largest peak.

A House 78

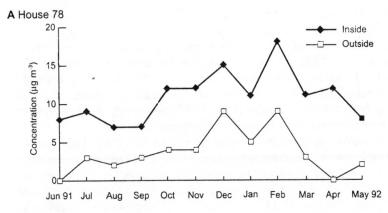

B House 49

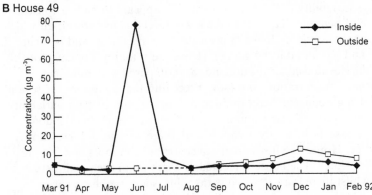

C House 116

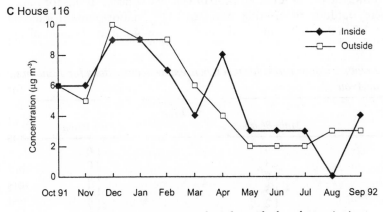

Figure 1   *Benzene concentrations inside and outside three homes in Avon over a 12 month period*

5.5.1 Temporal variation in concentrations. The relative standard deviation (RSD) of measurements of benzene for each house and room in the 174 homes is indicative of the amount of variability in concentrations of benzene between months. The mean RSD for measurements in the main bedrooms was 58% (range 0–199%) and this is lower than that found for TVOCs (67%) and higher than for formaldehyde (44%). This shows that benzene concentrations in the air of homes tend to be less variable than TVOC concentrations, indicating less influence of intermittent sources on concentrations.

Figure 1 shows the benzene concentrations in the main bedroom and outdoors for 3 homes that show varying amounts of within-year variability in concentrations indoors. It should be noted that readings below the detection limit of 2 μg m^{-3} are given a zero value. House 78 is an example of low inter-month variability indoors (RSD 31%), house 116 average variability (54%) and house 49 high variability (RSD 199%). House 78 appears to have a consistent indoor source of benzene that adds to the lower level due to outdoor air which shows a seasonal variation. House 78 is located in a rural area and is occupied by a smoker and has an attached garage. Hence the smoking and car probably account for the elevated concentration indoors compared with outdoors.

House 49 has concentrations outdoors exceeding those indoors for several months but a high concentration in June indoors results in high variability. Except for this peak, concentrations are higher outdoors than indoors in winter. This house is in a city location and there are no smokers and no attached garage. House 116 has similar concentrations indoors and out, again with a peak in winter. The house is located in a suburban area and there is no smoker and whilst there is an attached garage this is not used for a car.

Table 8 shows the indoor-outdoor ratio of benzene concentrations for the 13 homes where outdoor monitoring occurred. The indoor-outdoor ratio is greater than 1.0 in 9 of the homes and equal to 1.0 in 4 homes. In most of the

Table 8   *Indoor-outdoor ratios for mean benzene concentrations for 13 homes in Avon*

House Number	I/O Ratio	
	Main Bedroom	Living Room
2	0.9	1.0
20	2.3	2.0
27	1.0	1.0
33	1.3	1.3
44	1.2	1.7
49	1.8	2.3
60	1.3	0.8
66	1.3	2.0
78	2.8	3.0
96	1.2	1.2
116	1.0	1.2
128	1.0	1.0
140	1.3	1.3

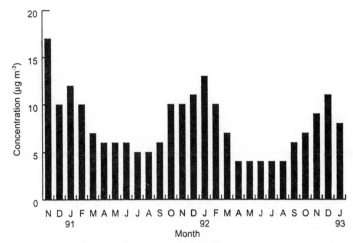

Figure 2   *Mean monthly benzene concentrations in main bedrooms for all homes*
($\mu g\ m^{-3}$)

homes the indoor-outdoor ratio for the main bedrooms and living rooms is similar. It should be noted that the outdoor value represents a reading at ground level in the rear garden of the house and depending upon proximity of major roads and any stratification of air, this may differ from the air entering the house via windows at various levels and locations and also by infiltration. The outdoor air as a source of benzene in the indoor air has a strong influence on the amount of intermonth variability in concentrations. This is in contrast to the situation for TVOC concentrations where the indoor-outdoor ratio is typically 10 rather than 1.4 as for benzene.

The influence of time of year and different years on benzene concentrations have also been examined. There was no statistically significant difference between the annual mean concentration measured in the main bedrooms of all homes during 1991 (7.8 $\mu g\ m^{-3}$) with that in 1992 (6.9 $\mu g\ m^{-3}$) (t-test, 95% confidence level). Figure 2 shows the mean benzene concentration in the main bedrooms for each month during the period November 1990 to January 1993. Highest concentrations occur in winter months and lowest in summer.

Figure 3 shows the mean concentration each month for all main bedrooms and outdoor sites monitored during the period June 1991-May 1992. This is the period for which the greatest amount of data was available; the minimum number of indoor measurements in any month was 62 and the minimum outdoor was 5. A strong seasonal variation is apparent and in the outdoor air the peak concentration in winter is 4 times higher than during spring and summer. This seasonal variation with highest concentrations in winter has also been shown by measurements of benzene using continuous monitors at 9 urban sites in England during 1994/95[35]. Table 9 shows mean concentrations for each season. In the outdoor air there was a statistically significant difference between all seasons except spring and summer (t-test, 95% confidence level). Indoors, concentrations in winter and autumn are significantly

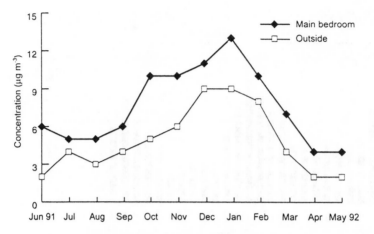

Figure 3   *Mean concentrations of benzene in main bedrooms and outdoors during the period June 1991 to May 1992*

higher than spring and summer. Also shown in Table 9 is the difference between benzene concentrations indoors and out for each season. This indicates that indoor sources of benzene have a greater influence on the indoor concentration during winter than summer months, probably due to lower rates of ventilation.

Table 9   *Mean benzene concentrations (all readings) in main bedrooms and outdoors for each season (June 1991–May 1992)*

Location	Benzene Concentration $\mu g\ m^{-3}$							
	Summer (June–August)		*Autumn (September–November)*		*Winter (December–February)*		*Spring (March–May)*	
	mean	*SD*	*mean*	*SD*	*mean*	*SD*	*mean*	*SD*
Outdoors (O)	3	2	5	3	8	2	3	3
Main Bedrooms (MB)	5	6	9	9	12	7	5	5
MB-O	2	–	4	–	4	–	2	–

5.6 Development of Sampling Strategies for Measurement of Benzene in Indoor Air

By investigating the temporal variation over a 12-month period it is possible to understand the quality of the data that would have been provided if a less intensive sampling strategy had been applied. Table 10 summarises the data for measurements of benzene in main bedrooms of 102 homes in the study for which 10 or more individual monthly readings were available. The full data is examined to show the effect of a reduced sampling frequency and the main outcomes are shown in Figures 4 and 5.

Table 10 *Effect of sampling frequency on the value of the measured annual mean benzene concentration for main bedrooms*

Sampling frequency	No of homes	Mean $\mu g\ m^{-3}$	SD	% deviation from mean annual benzene concentration in each home		
				mean deviation	minimum deviation	maximum deviation
Every month	102	7.5	1.2	0	0	0
Once a year						
(random selection)	102	7.0	6.9	35.2	0	141
January and July	102	8.0	4.7	21.9	0	102
Every other month (1,3,5,7,9,11th month)	102	7.6	3.8	7.4	0	55
Every other month (2,4,6,8,10,12th month)	102	7.4	3.8	7.5	0	55
Every 3rd month (1,4,7,10th month)	102	7.7	4.2	11.1	0	120
Every 3rd month (2,5,8,11th month)	102	7.2	3.4	9.5	0	51
Every 3rd month (3,6,9,12th month)	102	7.5	3.8	11.2	0	69
Every 4th month (1,5,9th month)	102	7.5	4.1	14.4	0	68
Every 4th month (2,6,10th month)	102	7.5	3.8	13.7	0	59
Every 4th month (3,7,11th month)	102	7.6	3.6	12.9	0	62
Every 4th month (4,8,12th month)	102	7.5	3.6	16.1	0	78
Twice a year (1,7th month)	102	7.8	3.8	21.3	0	102
Twice a year (2,8th month)	102	7.3	3.0	18.5	0	77
Twice a year (3,9th month)	102	7.6	3.9	17.1	0	72
Twice a year (4,10th month)	102	7.6	4.1	19.4	0	298
Twice a year (5,11th month)	102	7.2	3.6	16.6	0	62
Twice a year (6,12th month)	102	7.4	3.7	20.2	0	110

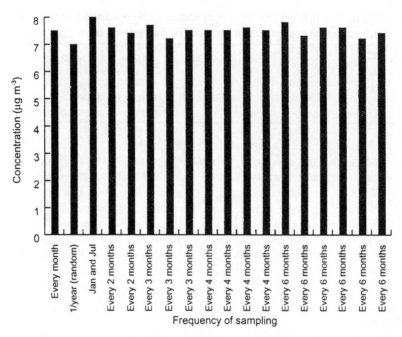

Figure 4   *Comparison of mean concentration of benzene in main bedrooms given by
various sampling strategies*

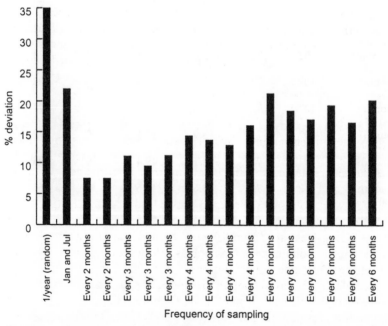

Figure 5   *Comparison of mean deviation from the annual mean concentration of
benzene resulting from the use of different sampling strategies*

Figure 4 illustrates the influence that a reduced sampling frequency has on the calculated annual mean concentration of the full population of houses in the study. All strategies gave a value within the range 7.0 to 8.0 µg m^{-3} which is within 7% of the actual value of 7.5 µg m^{-3}. Sampling once every 3 months gave values between 7.2 and 7.7 µg m^{-3}, a maximum deviation of 4%. This shows that, as with TVOC and formaldehyde[32], a much reduced sampling strategy than every month can give an accurate measure of the annual mean benzene concentration of a population of houses.

Figure 5 shows that to estimate the annual mean benzene concentration in any particular house, the uncertainty of the measurement is influenced more strongly by the sampling strategy. The use of one random sample in a 12-month period is least reliable, resulting in a mean deviation of 35% from the actual value. Sampling every 3 months gave a mean deviation of 9.5-11.2%. This is less than the deviation observed for measuring TVOC, which was 21-25% for sampling every 3 months, and for formaldehyde (13-15%). This reflects the lower within-year variability of concentrations of benzene compared with TVOC in particular. The required sampling strategy depends upon the objective of the measurement and the data from this study can be used to estimate likely errors associated with any particular strategy for measuring benzene in indoor air using diffusive samplers.

6 Conclusions

There are a wide range of indoor air pollutants that originate from both indoor and outdoor sources which can potentially have an adverse effect on the health and well being of building occupants. For some pollutants such as ozone the indoor environment is usually protective against polluted outdoor air, but for many others, including many volatile organic compounds, the concentrations indoors exceed those outdoors.

People spend 90% of their time in an indoor environment and it has been shown that personal exposure to a number of air pollutants is strongly influenced by the concentration in indoor environments where people spend most time. The exposure can also be enhanced by personal activities which result in higher concentrations in the breathing zone than the general indoor environment.

As part of its work to assess the risks to public health and well being due to air pollution in the home the UK Department of the Environment commissioned the Building Research Establishment to undertake a major study of some key indoor air pollutants in UK homes. Main findings from the study were that nitrogen dioxide concentrations were higher in homes with gas cooking, formaldehyde was higher in newer homes, volatile organic compounds were strongly affected by painting and decorating activities, geometric mean mite counts were about 180 mites m^{-2} carpet and fungi and bacteria counts in air were higher in summer than winter. Mean indoor concentrations of benzene were 1.3 times outdoor concentrations and outdoor concentrations were higher at city compared with rural sites. There was a statistically significant relationship between benzene concentrations and the presence of a smoker in the household and mean concentrations in homes with an attached garage were higher than those without an attached garage. There was a seasonal variation in benzene concentrations in indoor and outdoor air with highest concentrations occurring in winter months. An assessment of the error associated with determining the annual mean concentration using periodic sampling rather than continuous monitoring with diffusive samplers is presented to inform the planning of future studies of indoor air pollution.

7 Acknowledgement

The author thanks Miss V Brown for her contribution to the measurement of benzene in the BRE Indoor Environment Study and Mr S Chakrabarti and Mrs S Coward for statistical tests.

References

1. House of Commons. 'Indoor Pollution'. House of Commons Environment Committee, Sixth Report, HMSO, London, 1991.
2. Brown, V.M., Crump D.R., Dearling T.B., Gardiner D. and Gavin M.A. Proc of Healthy Buildings - IAQ '91, 325-328, Washington DC, September 1991.
3. Miller, E.R., Dearling T.B. and Boxall J. Interior painting of trim with solvent-borne paints. BRE Information Paper IP 8/95, BRE, Watford, UK, 1995.
4. Crump D.R. Issues in Environmental Technology, 4, 109-124, Royal Society of Chemistry, Cambridge, UK, 1995.
5. DOE, 'Urban Air Quality in the UK'. Department of the Environment, London, 1993.
6. Etkin D.S. 'Particulates in indoor environments'. Cutter Information Corporation, USA, 1994.
7. Brown V.M., Crump D.R. and Gardiner D. Indoor and ambient air quality, 423-430, Selper Ltd, London, 1988.
8. Schlitt H., Schauenberg H. and Knoppel H. *Proc. Indoor Air '93*, 2, 251-256 Helsinki, 1993.
9. Weschler C.J. and Fong K.L. Environment International 12, 93-97, 1986.
10. Brown S.K., Sim M.R., Abramson M.J. and Gray C.N. *Indoor Air*, 4(2), 123-134, 1994.
11. Maroni M., Seifert B. and Lindvall T. (ed). 'Indoor Air Quality', Chapter 1 Elsevier, Amsterdam, 1995.
12. Raw G. and Coward S. Proc Unhealthy Housing, University of Warwick, UK, December 1991.
13. Akland G.G., Hartwell T.D., Johnson T.R. and Whitmore R.W. *Environ. Sci. Tech.* 19(10), 911-918, 1985.
14. Clayton C.A., Perritt R.L., Pellizzari E.D., Thomas K.W. and Whitmore R.W. *J of Exposure Analysis and Env. Epidem*, 3(2), 227-250, 1993.
15. Dingle P., Hu S. and Murray F. *Proc of Indoor Air '93*, 2, 293-298, Helsinki, 1993.
16. Thomas K., Pellizzari E., Clayton C., Perritt L., Dietz R., Goodrich R., Nelson W.

and Wallace L. *J of Exposure Analysis and Environmental Epidemiology*, **3**(1), 49-73, 1993.

17. Wallace L., Pellizzari E. and Wendel C. *Indoor Air*, **4**, 465-477, 1991.
18. Crump D. Proc. Int. VOCs Conference, London, 235-246, November 1995.
19. Wallace L., Pellizzari E., Hartwell T., Davis V., Michael L. and Whitmore R. *Environment Research*, **50**, 37-55, 1989.
20. Maroni M., Seifert B. and Lindvall T. (ed). 'Indoor Air Quality'. Chapters 6-12, Elsevier, Amsterdam, 1995.
21. World Health Organisation. 'Air Quality Guidelines for Europe'. WHO, Copenhagen, 1987.
22. Kryzanowski M. *Proc. of Healthy Buildings '95*, **3**, 1841-1847, Milan, September, 1995.
23. Tobin R.S., Bourgeau R., Otson R. and Wood G.C. Proc. Indoor Air Quality, Ventilation and Energy Conservation, p. 12-26 Montreal, Canada, 1992.
24. Yoshizawa S. *Proc. of Indoor Air '96*, **1**, 21-28, Nagoya, Japan, July, 1996.
25. Maroni M., Seifert B. and Lindvall T. (eds) 'Indoor Air Quality', Chapter 37, Elsevier, Amsterdam, 1995.
26. CEC. 'Radon in Indoor Air'. European Collaborative Action on Indoor Air Quality, Report No 15, European Commission, Report EUR 16123 EN, Luxembourg, 1995.
27. CEC. 'Guidelines for Ventilation Requirements in Buildings. European Collaborative Action on Indoor Air Quality, Report No 11, European Commission report EUR 14449 EN, Luxembourg 1995
28. Dingle P. and Murray F. *Indoor Environ*, **2**, 217, 1993.
29. FISIAQ. 'Classification of Indoor Climate, Construction and Finishing Materials'. Finnish Society of Indoor Air Quality and Climate, Helsinki, June 1995.
30. The Institute for Environment and Health. IEH assessment of indoor air quality in the home; Nitrogen dioxide, formaldehyde, volatile organic compounds, house dust mites, fungi and bacteria. IEH Assessment A2. IEH, Leicester, UK, 1996.
31. Smith L. and Tuckett C. *Proc. of Indoor Air Quality and Climate*, **3**, 525-530, July, Nagoya, Japan, 1996.
32. Berry R., Brown V., Coward S., Crump D., Gavin M., Grimes C., Higham D., Hull A., Hunter C., Jeffrey I., Lea R., Llewellyn J. and Raw G. 'Indoor Air Quality in Homes - the BRE Indoor Environment Study. Building Research Establishment Report BR 299, Watford, UK, 1996.
33. Department of the Environment. 'Benzene', Expert Panel on Air Quality Standards, HMSO, London, 1994.
34. WHO. 'Indoor Air Quality : Organic Pollutants'. World Health Organisation, Euro report 111, Copenhagen 1989.
35. Bertorelli V. and Derwent R. 'Air Quality A to Z. Meteorological Office, Bracknell, UK, 1995.
36. Krause C., Mailahn W., Nael R., Schulz C., Seifert B. and Ullrich D. *Proc. Indoor Air '87*, **1**, 102–106, Berlin, 1987.
37. Wallace L. *Archives Environ. Health* **42**, 272-79, 1987.

Urban Air Pollution in the United Kingdom

R.M. Harrison

Institute of Public and Environmental Health, School of Chemistry, The University of Birmingham, Edgbaston, Birmingham B15 2TT, UK

1 Introduction

The 1990s have seen a burgeoning of interest in urban air pollution phenomena. This has arisen for two primary reasons. The first is that the majority of the population live in towns and cities and concern over adverse health effects of air pollutant exposure has increased greatly over this period, having been a matter of rather little public interest through the 1970s and early 1980s. Secondly, there has been a recognition that urban areas provide a distinct pollution climate which is rather different from that of the surrounding rural areas, and which may involve appreciably elevated pollutant concentrations, particularly during pollution episodes.

The distinct pollution climate of urban areas is associated predominantly with the high density of motor traffic, which provides a concentrated ground-level pollution source with a high impact on ground-level pollutant concentrations. Thus, a comparison of national emissions of particulate matter with emissions within London shown in Figure 1 serves to exemplify the far greater proportion of emissions arising from road traffic within London compared to the country as a whole. Since the road traffic emissions arise at ground level and are therefore not subject to the effective dispersion processes which influence elevated stack discharges before they reach the ground, the impact of urban road traffic emissions on air quality is very substantial. This should not be taken to imply that either road traffic or local emissions are the only influence on urban air quality. For example, in the case of airborne particulate matter, secondary particles which arise from the relatively slow atmospheric oxidation of sulfur and nitrogen oxides, and which hence affect both rural and urban areas rather similarly, are an important component of particulate matter pollution at urban sites, especially during the summer months. However, for other vehicle-generated primary pollutants such as carbon monoxide, oxides of nitrogen, benzene and for particulate matter in the winter months, local emissions from road traffic tend to dominate the urban pollution climate.

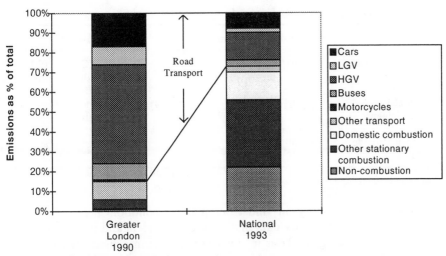

Figure 1 *Comparison of emission inventories for primary particulate matter for the UK and for London (from QUARG[1])*

2 Urban Monitoring Networks

Whilst nowadays the urban air pollution climate is dominated by emissions from road traffic, in the past smoke and sulfur dioxide emissions from coal combustion in domestic premises was far more important, and historically UK urban monitoring networks were designed to measure these pollutants. There remains a residue of the National Survey of Smoke and Sulfur Dioxide in which these pollutants are measured by techniques with a 24-hour averaging period. Such methods yield data only retrospectively after laboratory analysis and are hence unsuitable for real-time monitoring and provision of information to the public. In order to meet this need the Enhanced Urban Network of monitoring sites was established in the UK, the first site opening in 1992. This comprised high quality monitoring stations for continuous measurements of sulfur dioxide, oxides of nitrogen, ozone, carbon monoxide and particulate matter. More recently, this network has been augmented with other automatic monitoring stations, including those monitoring compliance with the European Union Nitrogen Dioxide Directive, and automatic stations established by local government. The combined network, which goes under the title of the Automatic Urban Network, now includes over 40 sites, with about 80 urban sites predicted for the end of 1996 (see Table 1). In addition, nitrogen dioxide is measured using inexpensive diffusion tube samplers which yield information on a longer timescale at some 1100 urban sites in a network operated by local government, but co-ordinated by NETCEN. These various activities are summarised in Table 1.

The Automatic Urban Network has led to a massive improvement in the quantity and quality of data available for public information and for use by

Table 1   *Summary of U.K. Urban Air Quality Networks* *

Name	Approx. no. of urban sites	Species measured
Automatic Urban	40	O_3, NO_x, CO, SO_2, PM_{10}
Hydrocarbon	11	Specific Hydrocarbons
Basic Urban	155	SO_2, Smoke
EC Directive	157	SO_2, Smoke
Multi-Element	5	Trace Metals
Lead	14	Lead
Nitrogen Dioxide	1100	NO_2 (Diffusion Tube)
TOMPS	3	Dioxins, PCB, PAN

* Site numbers at May 1996; subject to change

the research community in understanding air pollution processes. Some examples of how air quality data may be used to understand processes within the atmosphere are given later in this chapter. The measurement techniques used in the Automatic Urban Network are indicated in Table 2. Data are logged as 15 minute averages which are then combined into hourly averages which are passed by telemetry to a central control from which they are used for public information and entered into a database. Such data are described as provisional until they have passed through a full data ratification process.

Table 2   *Measurement techniques used in automatic networks*

Pollutant	Networks
Ozone	U.V. Absorption
Oxides of Nitrogen	Chemiluminescence
Carbon Monoxide	I.R. Absorption
Sulfur Dioxide	U.V. Fluorescence
PM_{10}*	Tapered Element Oscillating Microbalance
Hydrocarbons	Cyclic Gas Chromatography

* PM_{10} is particulate matter passing an inlet with a 50% cut-off efficiency at 10 μm

3 Urban Pollution Episodes

For the primary air pollutants there are two main controls on airborne concentrations. The first is the source strength. Thus, traffic-generated pollutants typically show a diurnal fluctuation exemplified by Figure 2, the main controlling factor being volumes of road traffic. Morning and evening rush hour peaks are clearly visible with a minimum typically occurring about 4-5 am when road traffic is at its lightest. The other major control upon air pollutant concentrations is provided by meteorology. This determines both the rate and extent to which the pollutants mix upwards from their ground-level source and are thereby diluted by vertical movement. Additionally, horizontal winds dilute pollutant emissions and carry them out of urban areas and horizontal windspeeds are very important determinant of air pollutant concen-

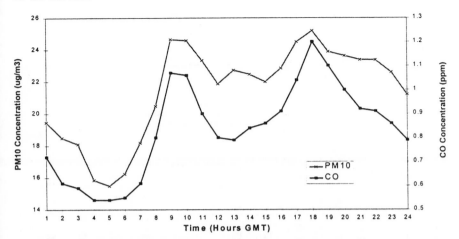

Figure 2   *Average diurnal variation in carbon monoxide and particulate matter, Birmingham 1993–94*

trations. Occasionally, particularly in winter, stagnant conditions prevail for a number of consecutive days. During such occasions, horizontal windspeeds fall to near zero and vertical dispersion of pollutants is severely limited by the presence of a low level temperature inversion which provides an effective lid to the atmospheric boundary layer. Under these conditions, severe pollutant build-up can occur and in December 1952, over a period of some four days, concentrations of smoke and sulfur dioxide built up massively resulting in an estimated 4000 excess deaths. Fortunately, pollution control policies have greatly reduced the severity of air pollution episodes, and in a period of stagnant weather in December 1991 in London, it was concentrations of the

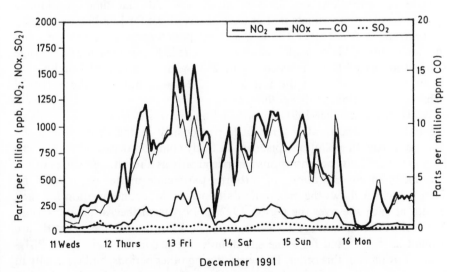

Figure 3   *Concentrations of continuously measured pollutants in London in the December 1991 episode*

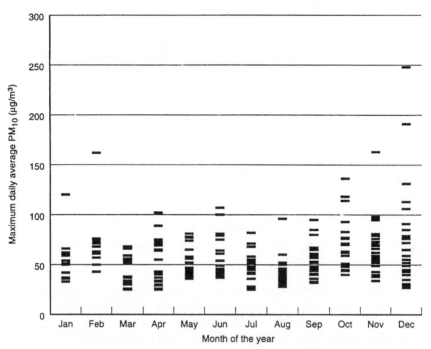

Figure 4   *Maximum daily mean PM_{10} concentration for each month at each site in the automatic urban network, 1992-94 (385 site months) (from QUARG[1])*

traffic-generated pollutants, carbon monoxide, oxides of nitrogen and hydro-carbons which built up to very high concentrations (see Figure 3) whilst sulfur dioxide concentrations remained relatively low. At that time, continuous measurements of particulate matter were not available for London, but scaling concentrations from other traffic-generated pollutants and estimation from measurements of black smoke made at the time indicates that hourly average concentrations of PM_{10} reached about 250 $\mu g \ m^{-3}$ in central London, far below those prevailing in the London 1952 episode, but well above typical urban concentrations of this pollutant.

Whilst the December 1991 episode in London was exceptional in terms of the pollutant concentrations reached, particularly in the case of nitrogen dioxide, urban air pollution episodes are fairly common occurrences, particularly in winter, and concentrations can rise greatly as a result. Some idea of the episodicity in concentrations of airborne particulate matter is provided by Figure 4 which shows the maximum daily mean PM_{10} concentration for each month at each site over the period 1992 to 1994 from the Automatic Urban Network. Annual average concentrations of PM_{10} across the network are around 25 g m^{-3} and for some site months the maximum daily mean PM_{10} concentration is of this order. However, during other periods, and especially in the winter months, far higher concentrations are encountered with maximum daily average PM_{10} reaching almost 250 $\mu g \ m^{-3}$ during a severe episode.

4 Trends in Emissions and Urban Air Quality

Two factors need to be considered to understand past and present trends in air quality. The first relates to low level emissions of smoke and sulfur dioxide from burning coal in domestic premises. As mentioned above, this source once dominated the UK urban air pollution climate and led to very serious pollution episodes in the 1950s and 1960s. As the smoke and sulfur dioxide source has diminished, so emissions from road traffic have increased. The extent of growth in the latter source has been determined by the balance of two factors which operate in different directions. Until recently the dominant factor was the growth in road vehicle traffic which led to an increase in emissions of most traffic-generated pollutants up until 1992[3]. This growth was counteracted to some degree by improved vehicle emission standards which led to a decrease in pollutant emissions per vehicle-kilometre travelled. The requirement to fit catalytic converters to all new petrol cars from January 1993 caused a dramatic improvement in emissions per vehicle (see Figure 5) and this factor is now dominant, leading to an anticipated decline in emission of traffic-generated pollutants for the next few years. This tendency is complicated, however, by an increase in the market share of diesel cars which took place in the early 1990s, which will have a beneficial effect on emissions of carbon monoxide and most hydrocarbons, but will lead to a lesser improvement in emissions of oxides of nitrogen and particulate matter than would have occurred for a wholly petrol-powered fleet[5] (see Figure 5).

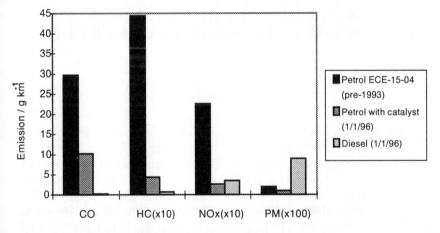

Figure 5 *Comparison of typical exhaust emissions from pre-catalyst petrol car, petrol car with three-way catalyst and diesel car (data from ETSU[4])*

There is a long historical dataset for urban concentrations of smoke and sulfur dioxide exemplified by Figure 6. A change from bituminous coal to cleaner fuels such as gas for home heating has had a tremendous beneficial effect on the quality of urban air. Nowadays, high level emissions from power

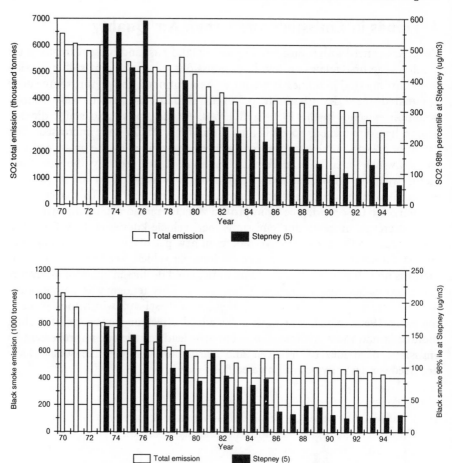

Figure 6   *Concentrations of (a) sulfur dioxide and (b) smoke in air in London in relation to emissions from 1970-1995*

stations are the major source of sulfur dioxide, and urban and rural concentrations have converged to the point where there is little difference. The UK is committed by international agreements to further reductions in sulfur dioxide emissions and concentrations will continue to fall. In contrast, whereas emissions of smoke from coal combustion have fallen greatly, emissions of smoke particles from diesel vehicles have increased[3], and this is now the major source. Thus concentrations of black smoke in urban air have tended to level off (Figure 6).

A largely vehicle-generated pollutant, which has been subjected to highly effective control, is lead. Lead was for many years universally added to petrol as an octane enhancer and was emitted to the atmosphere as fine particles. Figure 7 illustrates how changes in the quantities of lead used in petrol have been reflected in improved air quality in central London. At the end of 1985 a sharp decrease occurred due to a reduction in the maximum permitted lead

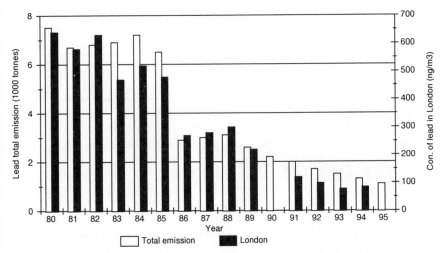

Figure 7   *Concentrations of lead in air in central London and national use of lead in petrol from 1980-1995*

content of petrol from 0.40 g l^{-1} to 0.15 g l^{-1}. More recently, an increasing market penetration of unleaded fuel has led to a continued reduction in emissions. This trend is certain to continue due to the requirement of unleaded fuel for new cars, and a proposal by the European Commission to phase out leaded fuel entirely by 2000.

UK emissions of oxides of nitrogen peaked in 1989, and emissions from the road traffic sector in 1992[3]. Since that time, catalytic converters have led to a reduction in NO_x emissions and the first signs of some reduction in NO_x concentrations at urban sites are now evident. NO_x emissions from road vehicles are mainly in the form of nitric oxide (NO), the less toxic of the two compounds making up NO_x. NO is oxidised in the atmosphere to the more toxic nitrogen dioxide which is the compound for which air quality standards have been set. The relationship between concentrations of NO_x and NO_2 is a complex one, as exemplified by Figure 8 which shows hourly average concentrations measured in winter at a central London site (see also next section).

Over a very wide range of NO_x concentrations, the NO_2 concentration is rather insensitive to changes in NO_x (the flat portion of the graph). Because of this, annual average concentrations of NO_2 are unlikely to change radically as a result of changing NO_x emissions (and atmospheric concentrations). However, the implication of the steepness of the upper part of the curve is that modest reductions in NO_x should lead to substantial reductions in peak nitrogen dioxide concentrations. The prognosis is therefore rather good, although to achieve a situation whereby the World Health Organisation draft revised air quality guideline for nitrogen dioxide of 110 ppb hourly average concentration is not exceeded, would require an approximate halving of nitrogen oxide emissions, which is unlikely to occur in the foreseeable future.

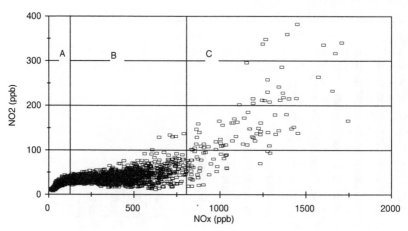

Figure 8   *Relationship of hourly NO_2 and NO_x at Cromwell Road, London in winter months (from Harrison and Shi[6])*

The two hydrocarbons for which the Expert Panel on Air Quality Standards (EPAQS) has recommended air quality standards are benzene and 1,3-butadiene, both of which have carcinogenic activity. These are both primary pollutants present in vehicle exhaust and both are pollutants whose emission is reduced effectively by the use of catalytic converters. Thus, as shown in Figure 9 for benzene, the predicted future prospects with respect to benzene concentrations are good. Currently annual average benzene concentrations at background sites in UK urban areas are around 1-2 parts per billion (ppb) which lies below the limit recommended by EPAQS of 5 ppb, and close to the long

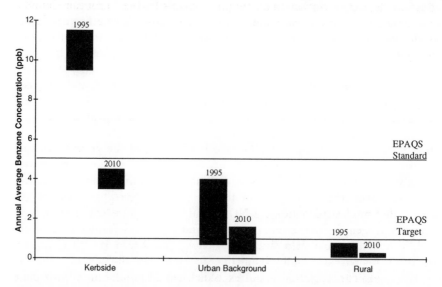

Figure 9   *Predicted changes in benzene in air concentrations at UK sites between 1995 and 2010 (from Derwent[7])*

term target of 1 ppb which EPAQS also recommended. Substantial reductions in 1,3-butadiene concentrations can also be anticipated.

The two air pollutants with the most obvious effect on public health in the UK are particulate matter and ozone. Both are complex in terms of their sources and their control. Particulate matter currently measured in the UK as PM_{10} has both a primary component, described by the inventories in Figure 1, and a secondary component which is formed within the atmosphere. The secondary component, whose role is most prominent in the summer months, is formed within the atmosphere from the oxidation of gaseous sulfur and nitrogen oxides. This component can thus be controlled by limiting emissions of the precursor gases and it is estimated that emissions reductions currently agreed will lead to a reduction of approximately 40% in secondary particle concentrations in the UK atmosphere by the year 2005[1]. Primary emissions, which are dominant in winter, and most important in urban pollution episodes, arise mainly from road traffic. Motor vehicles are subject to increasingly strict standards determined by Europe, but it is clear from Figure 5 that because of the very substantial differences in particulate matter emissions from petrol and diesel cars, the future air pollution by PM_{10} will depend critically upon the relative market share of the two fuel types in the car market. QUARG[1] estimated that to achieve the EPAQS recommended standard for PM_{10} of 50 μg m^{-3} as a 24-hour running average at city centre background sites, a reduction of approximately two thirds in road traffic emissions of particulate matter would be required. If roadside locations and severe episodes were to be included, a reduction of the order of 80% would be necessary. Projections based upon currently agreed emissions standards for road vehicles indicate a reduction of approximately 50% by the year 2010, which will clearly not be sufficient. However, further tightening of controls by the European Commission is anticipated in the meantime and it may be possible to achieve sufficient control that the standard be met. However, this outcome is crucially dependent on the mix of vehicles in the car market, as illustrated by Figure 10, which

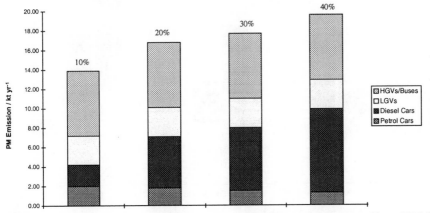

Figure 10 *Estimated emissions of particulate matter from urban road traffic in 2005 for different percentage diesel cars in the fleet (from ETSU[4])*

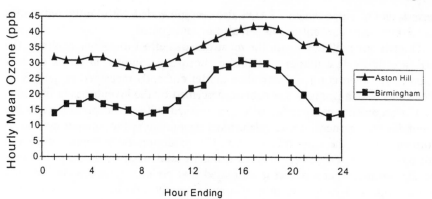

Figure 11 *Comparison of average summer diurnal profiles of ozone in air at an urban site (Birmingham) and a rural site (Aston Hill) in the same part of the UK (from QUARG[8])*

shows the sensitivity of particle emissions from the car sector and from road traffic as a whole to the uptake of diesels in the car market.

The other pollutant for which prediction is very difficult is ozone. This is entirely a secondary pollutant formed within the lower atmosphere by reactions of nitrogen oxides and hydrocarbons in sunlight (see also next section). Because of the complex chemistry of ozone, urban concentrations are typically lower than rural levels (see Figure 11) due to destruction of ozone by reaction with fresh nitrogen oxide emissions from road traffic within city centres[8].

One consequence of the reduction in nitrogen oxide emissions from road

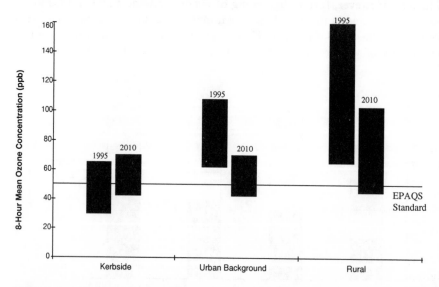

Figure 12 *Predicted changes in ozone in ground-level air at UK sites between 1995 and 2010 (from Derwent[7])*

traffic is that the urban and rural curves exemplified by Figure 11 will tend to converge such that rural concentrations will fall, but urban concentrations may well rise in the short term before the fall in rural levels is of sufficient magnitude to be dominant. Prediction of future changes in urban ozone is extremely difficult since these changes and their geographic distribution are critically dependent on the relative reductions across Europe in oxides of nitrogen and hydrocarbons, the substances contributing to ozone formation. Additionally, rather different pictures emerge if annual average ozone concentrations and peak concentrations are considered. However, Figure 12 portrays predicted changes in ozone at the year 2010 afforded by the emission control policies currently in place.

5 Chemistry of Urban Air Pollution

The timescale of residence of air pollutants in the urban atmosphere is relatively short; it is rarely more than a few hours and is frequently an hour or less. Thus, atmospheric chemical processes such as the oxidation of sulfur dioxide, which are important on a regional scale, have little influence on atmospheric pollutant composition on the scale of a UK urban area. Secondary pollutants such as sulfate, which are formed only rather slowly in the atmosphere, show rather uniform concentration fields with little difference between urban sites and the surrounding rural area.

Some atmospheric chemical processes are, however, sufficiently rapid to have major influences on the scale of an urban area. The most notable of these relate to the $NO/NO_2/O_3$ system which is reactive on a timescale of seconds. The three rapid reaction processes of greatest importance are the following:

$$NO_2 + h\nu \quad \rightarrow \quad NO + O \tag{1}$$
$$O + O_2 \quad \rightarrow \quad O_3 \tag{2}$$
$$NO + O_3 \quad \rightarrow \quad NO_2 + O_2 \tag{3}$$

In rural areas, additional chemistry based on peroxy radicals formed from reactions of hydrocarbons is important in converting NO to NO_2 without consuming ozone. This can lead to the build-up of enhanced concentrations of ozone which are widely observed in rural areas of North-Western Europe during sunny summer anticyclonic weather conditions. When such ozone-polluted air is advected into urban areas, it encounters freshly emitted NO from vehicle exhausts (vehicle-emitted NO_x is predominantly in the form of NO). This leads to a local destruction of ozone (equation 3) which is seen clearly in Figure 11 which compares seasonally averaged diurnal concentration profiles of ozone at rural and urban sites in the same part of the country. Clearly, the central urban sites have on average a much suppressed concentration of ozone.

It is ozone which is predominantly responsible for the conversion of NO, emitted from combustion sources such as motor cars, into NO_2, the more toxic of the oxides of nitrogen. This is primarily by reaction (3) above, but the implication of this reaction is that the concentration of nitrogen dioxide, which can be formed as a result of it, cannot exceed the concentration of ozone available to carry out the oxidation. In general, this relationship holds and nitrogen dioxide concentrations rarely exceed concentrations of ozone in the upwind background air. However, there are two circumstances under which relatively high concentrations from nitrogen dioxide can form. As mentioned above, in summer, reactions involving peroxy radicals formed from hydrocarbons in the atmosphere can cause conversion of NO to NO_2 without consuming ozone, thus leading to elevated NO_2 concentrations in summer episodes. Yet higher nitrogen dioxide concentrations have been observed in winter episodes, most notably the December 1991 London episode (Figure 3) described above, during which the concentration of nitrogen dioxide at Bridge Place in central London reached 423 ppb, which vastly exceeds the concentration of ozone in background air available to oxidise NO. The relationship between hourly average NO_2 and NO_x is shown in Figure 8 from data taken from a site in central London. This shows an interesting pattern of behaviour in which, at very low NO_x concentrations, NO_2 increases with NO_x as the net effect of the three reactions above is to oxidise a large proportion of NO to NO_2. The curve then enters a very flat portion in which there is little increase in NO_2 as NO_x increases. This is the ozone-limited region within which the nitrogen dioxide concentration is determined by the ozone concentration in upwind background air, plus a contribution from the small proportion of NO_2 in the NO_x emissions from road traffic (typically 5%). Then lies the most interesting part of the curve in which NO_2 rises rapidly for increasing NO_x. Detailed analysis of these high NO_x conditions indicates that reaction (4) plays a role:

$$2NO + O_2 \quad \rightarrow \quad 2NO_2 \tag{4}$$

This reaction is second order in NO concentration and plays no role until NO exceeds about 500 ppb. Detailed analysis indicates that even accounting for the contribution of this reaction, the above chemistry cannot account for the rate of production of nitrogen dioxide in severe pollution episodes and further chemical processes entailing dark reactions of unsaturated hydrocarbons has been invoked to explain NO_2 formation rates[6].

6 Urban Air Quality Management

With the improved availability of air quality data and the development of ambient air quality standards, attention has turned in the UK to the management of urban air quality. Figure 13 illustrates some of the key facets of the urban air quality management process.

Monitoring data are central to urban air quality management as they provide evidence of the current state of the atmosphere. These data must be

THE URBAN AIR QUALITY MANAGEMENT PROCESS

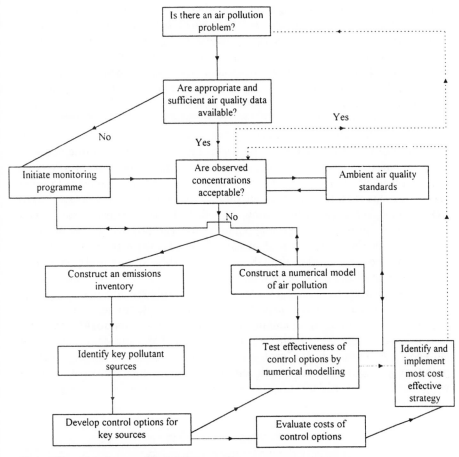

Figure 13 *Key facets of the urban air quality management process*

judged against some measure of acceptable air quality and this usually takes
the form of an air quality standard or objective. For some years the UK has
been required to meet air quality standards set by the European Union. These
standards are based upon evidence of effects on human health embodied in the
World Health Organisation Air Quality Guidelines for Europe, translated by
the Commission into limit and guide values which apply across the European
Union. There is a legal requirement under European law to comply with EU
Directive limit values which currently exist for sulfur dioxide and smoke,
nitrogen dioxide and lead. Ozone is the subject of a further Directive, which
does not contain a strict limit value. More recently, the United Kingdom has
embarked on setting its own air quality standards. These are recommended by
a committee known as the Expert Panel on Air Quality Standards (EPAQS)
which reports to the Department of Environment. Thus far, EPAQS has

recommended standards for ozone, benzene, 1,3-butadiene, carbon monoxide, sulfur dioxide and PM_{10}. Recommendations will also be made in respect of nitrogen dioxide, lead and polycyclic aromatic hydrocarbons. The UK government has proposed an air quality strategy in which these standards, with a small percentage of permitted exceedences for ozone, PM_{10} and sulfur dioxide, will be regarded as objectives to be met by the year 2005[9]. In the meantime, the EPAQS and EU standards provide excellent benchmarks against which to judge the acceptability of air quality. If this is found to be unsatisfactory due to exceedence of EPAQS standards, then remedial action should be contemplated.

Remedial action to improve air quality in urban areas is inevitably expensive. Since many urban pollution hotspots are caused by road traffic emissions, prediction of the effect of pollution control measures such as traffic management schemes is highly complex, since diverting traffic away from a hotspot area will lead to enhanced traffic and possible air pollution problems in other areas. For this reason, numerical air quality management models (see Figure 13) are being developed which allow a prediction of the effect of air quality management strategies upon air pollution concentrations to be made in advance of the implementation of such strategies. Whilst such models have yet to be deployed and tested in the UK, their use will be an essential prerequisite to cost-effective local management of air quality.

Pollutant emissions from the vehicle fleet are highly sensitive to the age and state of maintenance of the vehicles. Figure 14 shows, for measurements in Leicester, the relationship of exhaust emissions of carbon monoxide measured on the road to the age of the vehicle (x-axis) divided into five quintiles by emission concentration for each vehicle age. This shows clearly that (i) pollutant emissions increase with the age of the vehicle, and (ii) a small proportion of the vehicles is responsible for a large proportion of the pollutant emission for all vehicle ages.

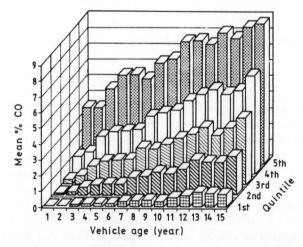

Figure 14   *Exhaust gas concentrations of carbon monoxide measured on the road in Leicester by vehicle age, divided into quintiles (from Zhang et al.[10])*

The percentage of the vehicles responsible for 50% of the emissions from the entire fleet is termed the percentage "gross polluters" and in one study[10] this was found to be 18.2% in Leicester; 13.4% in Edinburgh, and only 8.4% in London, where the car fleet is generally better maintained. This survey was carried out in 1992 when very few catalyst-equipped vehicles were on the road. Studies in other countries where catalysts were adopted earlier tend to show a step-change between cleaner catalyst-equipped cars and others, but still show a deterioration with age of the catalyst-equipped fleet and the presence of "gross polluters" with defective catalysts. Clearly, vehicle inspection and maintenance programmes have a major part to play in minimising pollutant emissions.

7 Concluding Remarks

The high level of public and media concern over air pollution in recent years might lead the uninformed to the view that air quality must be deteriorating. Fortunately, the reality is rather different and for most urban pollutants the advent of cleaner road vehicles is leading to a steady reduction in air pollutant concentrations. There is a growing awareness amongst the regulators that further controls on road vehicle emissions will become increasingly expensive, and cost-benefit studies are being conducted to evaluate the most cost-effective strategies for control. The European Auto-Oil Study is an example of a programme in which numerical models of air quality have been devised for a number of European cities and control strategies involving improved technology for vehicles, and cleaner fuels have been evaluated. Subsequent costings of the various options have assisted the European Commission in proposing vehicle emission and fuel quality standards for the years 2000 and 2005. Increasingly, as incremental improvements in air quality become more expensive, this kind of approach will be an essential component of the strategic control of air pollution. However, provided traffic growth is not allowed to wipe out the benefits of cleaner vehicles, the outlook for urban air quality is good.

References

1. Quality of Urban Air Review Group, 'Airborne Particulate Matter in the United Kingdom', Department of the Environment, London, 1996.
2. J.S. Bower, G.F.J. Broughton, J.R. Stedman and M.L. Williams, *Atmos. Environ.*, 1994, **28**, 461-476.
3. Department of Environment, 'Digest of Environmental Statistics', HMSO, London, No. 18, 1996.
4. ETSU, 'U.K. Petrol and Diesel Demand', ETSU, Harwell, 1994.
5. Quality of Urban Air Review Group, 'Diesel Vehicle Emissions and Urban Air Quality', Department of Environment, London, 1993.
6. R.M. Harrison and J.P. Shi. *Sci. Total Environ.*, 1996, **189/190**, 391-399.
7. R.G. Derwent. *Clean Air*, 1995, **25**, 70-94.

8. Quality of Urban Air Review Group, 'Urban Air Quality in the United Kingdom', 1993.
9. Department of Environment, 'The United Kingdom National Air Quality Strategy: Consultation Draft', Department of Environment, London, 1996.
10. Y. Zhang, D.H. Stedman, G.A. Bishop, P.L. Guenther, S.P. Beaton. *Environ. Sci. Technol.*, 1995, **29**, 2286-2294.

Rural Air Pollution in the United Kingdom

[1] D. Fowler, R. Smith, M. Coyle and M. Sutton
[2] G. Campbell, C. Downing and K. Vincent

[1] Institute of Terrestrial Ecology Edinburgh, Bush Estate, Penicuik, EH26 0QB, UK
[2] National Environmental Technology Centre, Culham, Abingdon, OX14 3DB, UK

1 Introduction

Changes in rural air pollutant problems in the UK are a reflection of some of the major changes in social and industrial activity. At the turn of the century air quality in the UK was particularly poor in the large cities as a consequence of domestic and industrial combustion, and was good in rural areas and at the coast. Some coastal resorts even provided simple measurements of ambient ozone and promoted this as a measure of the purity of the air. Relative to air quality in the large cities these claims were reasonable but as will be explained later, ground level ozone is no longer an appropriate indicator of good air quality. In fact ground level ozone has become the most important air pollutant in rural UK and may be the cause of significant crop loss and human health problems for the UK population.

The change from urban dominated pollutant sources and problems to regional and long-range issues was gradual and has been complicated by major changes in the nature and distribution of sources as well as the chemical form of the dominant pollutants. It is also clear from the recent measurements of air quality in urban areas, that although the composition of the pollutant mixture has changed as motor vehicles have taken over from domestic coal burning and industry as the dominant sources, air pollution in cities remains a major problem.

This review is confined to the air pollutants in rural areas. The focus of the paper is on the current UK rural air pollutants, their distribution and fate and recent changes in emission and deposition. For convenience the paper considers each of the major pollutants in turn, sulfur, oxidized and reduced nitrogen and photochemical oxidants. There are significant emissions of heavy metals, persistent organic pollutants (POPs) and HCl but as minor components of current emissions these have been omitted from this paper.

Table 1   *Environmental issues*

Pollutants	Problems	Targets
NO_x, SO_2, NH_x	Acid deposition	Freshwater Soils Materials
NO_x, NH_x	Eutrophication	Natural ecosystems Human health
NO_x, VOCs	Photochemical oxidants	Climate change Crop loss Ecosystems

2 Environmental Issues

As there are important overlaps between the emitted pollutants and their environmental effects, it is worth briefly summarizing the form of the emitted pollutants and the environmental problems with which they are associated.

2.1 Acidic Deposition

The emitted pollutants contributing to acidic deposition include SO_2, NO, NO_2 and NH_3. The emission of NH_3 contributes to acidification following deposition, uptake and assimilation by plants. The assimilated form $R\text{-}NH_2$ effectively releases hydrogen ions to the soil. The sensitive components of terrestrial ecosystems include poorly buffered freshwaters and soils, but calcareous building materials are also very sensitive.

2.2 Ecosystem Eutrophication

The atmospheric inputs of oxidized and reduced nitrogen to nitrogen sensitive ecosystems may lead to substantial changes in species composition and carbon sequestration.

2.3 Photochemical Oxidants

Emissions of volatile organic compounds (VOC) and oxides of nitrogen provide the reactants for the production of tropospheric ozone, peroxyacetyl nitrate (PAN) and related photochemical oxidants. The environmental targets include agricultural crops, forests, human health and materials.

The oxides of nitrogen play a central role in each of these three pollutant problems (Table 1) and for which control measures are currently being developed.

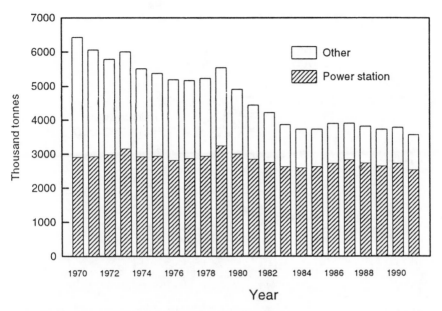

Figure 1   *Estimated emissions of SO₂, 1970-1991*

3 Sulfur

Emissions in the UK of sulfur as SO_2 peaked in 1970 at just over 3×10^6 tonnes SO_2-S and have declined by about 50% to approx 1.4×10^6 tonnes SO_2-S in 1994 (Figure 1). The decline in emissions includes a substantial switch of UK industry from coal to gas, but also includes reduced domestic coal use and more recently the use of flux-gas-desulfurization (FGD) equipment by major UK power companies.

During the late 1970s and early 1980s a network of primary and secondary precipitation chemistry collectors was established in the UK. This provided annual wet deposition maps for the UK from mid 1980. It was recognised that the high rainfall areas of the uplands of western UK were particularly difficult to monitor, because the locations receiving the largest inputs (which are among the most acid-sensitive regions) are inaccessible and a substantial fraction of the annual wet deposition occurs as snow. These regions receive large inputs through orographic effects in which frontal precipitation washes out polluted orographic cloud as illustrated schematically in Figure 2. The orographic effects on wet deposition have been incorporated into procedures to map sulfur deposition throughout the country. The total quantity of sulfur wet deposited in the mid 1980s was typically 230 kT-S or about 12% of emissions, averaging 10 kg S ha^{-1} throughout the country.

Wet deposition is relatively easily measured using networks of simple collectors. Dry deposition however, is notoriously difficult to measure and

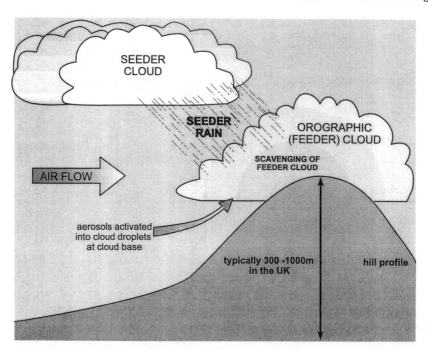

Figure 2   *The seeder-feeder process*

there are no networks of dry deposition monitoring stations. In the UK a network of rural SO_2 concentration monitoring stations was established during the 1980s to complement a long-running network of smoke and SO_2 monitoring sites largely in cities but including a limited number of suburban and rural sites. The rural SO_2 measurements were used to define the spatial patterns and seasonal variability in rural concentration. Dry deposition was estimated using a process based model[1] using the SO_2 concentration map. The key parameters for the model were quantified in field measurements using micrometeorological methods, combined with meteorological and land use data to generate annual inputs[2]. The resulting maps were therefore measurement based but only validated over short periods and land use categories. The annual dry deposition was estimated at 240 kT SO_2-S for 1987 using these methods. The sulfur budget for the atmosphere over the UK could therefore be calculated directly from the wet and dry deposition fields and compared with emissions, which were quite well established. The resulting budget (Figure 3a) shows that 25% of UK emissions were deposited within the UK land surface area averaging 20 kg S ha^{-1} and ranging from 5 to 100 kg S ha^{-1}. The spatial resolution scale of the wet and dry deposition maps was 20 km × 20 km and was limited mainly by uncertainties in the concentration fields of both gaseous SO_2 in source regions and SO_4^{2-} in precipitation in complex terrain.

Between the mid 1980s and mid 1990s emissions of SO_2 declined by 28% from 1900 kT S to 1350 kT S. During this period the dry deposition estimates

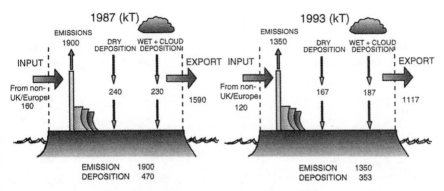

Figure 3   *The UK atmospheric sulfur budget for 1987 and 1993*

have been improved by operating at 2 locations a continuous measure of SO_2 dry deposition using micrometeorological methods. One flux monitoring station is sited in a large arable field in rural Nottinghamshire within the region of largest SO_2 concentration (4 ppb), and the other is operating at a moorland site in the Scottish borders at which SO_2 concentration averages 0.2 ppb. These flux monitoring stations yield between 3000 and 6000 hours of flux measurements each year and provide the necessary data to quantify the components of canopy resistance for SO_2 uptake on wet and dry cuticular surfaces, soil as well as stomatal uptake[2].

The field data for Sutton Bonnington may be compared with the dry deposition model for dry (Figure 4a) and wet days (Figure 4b) and for a wide range of other surface and atmospheric conditions. The data show that wet surfaces are a major sink for dry deposition and that in the East Midlands terrestrial surfaces are wet with rain or dew for typically 60% of the time. The canopy resistance for wet surfaces is small but non-zero, with values generally in the range 10 S m^{-1} to 30 sm^{-1}. Thus these surfaces are not 'perfect sinks' for SO_2 as had been reported earlier[3]. The application of zero canopy resistance for wet surfaces in national mapping of dry deposition leads to very large over-estimates in the dry deposition field. In the UK, the rate of dry deposition expressed as a deposition velocity (Vg) is typically 5 to 10 mm s^{-1}.

The UK and the Netherlands are the only countries in which rates of dry deposition are routinely monitored. Other countries (Germany and USA) operate inferential deposition stations at which the ambient concentration of SO_2 is monitored and an associated weather station provides the atmospheric components of the resistance network. However, the inferential approach does not provide any direct information on the affinity of the absorbing surface for SO_2 and thus the value of r_c, the canopy resistance which is generally the dominant term (Figure 5). Alternative strategies for estimates of dry deposition include throughfall measurements below canopies of trees[4]. Such methods are helpful integrators of long term flux, but must be calibrated with a direct measurement of the flux to avoid important systematic errors. Thus the UK, with a modest network of rural SO_2 monitoring stations and intensive direct

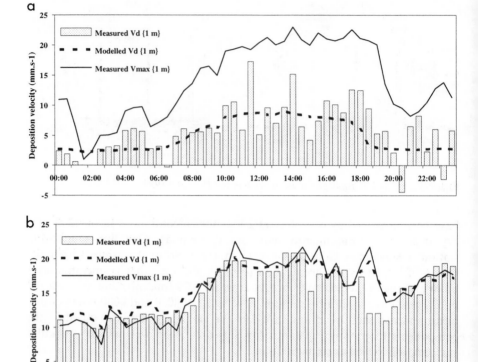

Figure 4   *Measured and modelled dry deposition velocity of SO₂ onto (a) a dry wheat canopy (April 1993) and (b) a wet sugar beet canopy (September 1993)*

measurements of dry deposition at 2 sites has a direct method for validating the dry deposition model used unlike most other countries. The uncertainty in absolute values for an individual 20 km × 20 km grid square is large ($\pm$ 50%), mainly as a consequence of uncertainty in the SO_2 concentration field.

The most important application of the dry and wet deposition monitoring networks is to provide maps of deposition against which an assessment of effects may be made. The acidification effects are quantified using a critical loads (CL) approach, and grid squares which exceed CL are identified. In the UK the majority of the soils and freshwaters which exceed CL for acidification are in the uplands of the west and north. Deposition inputs in these areas are dominated by wet deposition.

3.1 Recent Trends in Sulfur Deposition in the UK

In the period 1987 to 1993 emissions of sulfur declined by approximately 25% (Figures 3a and 3b). The partitioning between wet and dry deposition changed

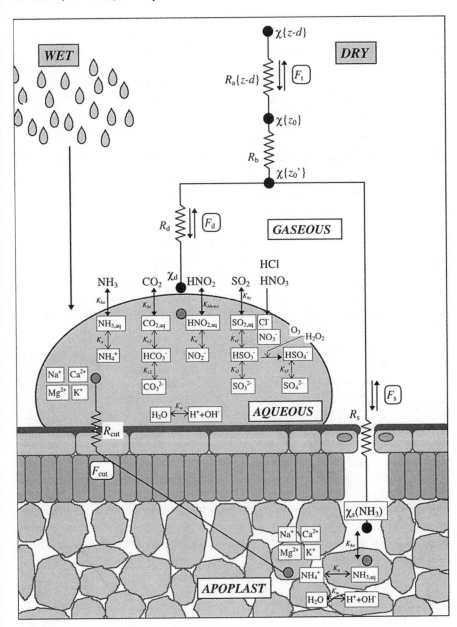

Figure 5 *A deposition model*

during the period with a decline in wet deposition of 19% while the decline in dry deposition was 31%. Thus the benefits of reduced emissions have been primarily in dry deposition, which occurs mainly close to the source regions, where SO_2 concentrations are largest. The high wet deposition regions which include most of the areas of exceedance showed an improvement, and some grid squares no longer exceeded the CL threshold. However, total deposition in these areas declined least. The longer term consequence of the changes in partitioning wet and dry deposition are of much greater concern. Because the underlying processes which have caused this change are unknown, we cannot be confident in long-term projections of CL exceedance (e.g. for 2005). The observed changes in wet and dry deposition provide strong evidence that the relationship between emissions and deposition are non-linear. Thus sulfur deposition at any point in the UK landscape may not decline linearly as UK emissions decline.

The causes of the observed non-linearity are unknown at present but those most likely include:

a) Oxidant limited conversion of SO_2 to $SO_4{}^{2-}$.
b) A change in the source height of emission during the period of study.
c) A change in the dry deposition velocity with time (deposition velocity increasing with time).

Of the three possible mechanisms, the absolute change in low level sources seems to be too small over the 8 years to explain observed changes in dry deposition. The oxidant-limited SO_2 to $SO_4{}^{2-}$ conversion might be expected since in the large plumes of SO_2, the supply of oxidants limits rates of oxidation. The change in deposition velocity also seems plausible, since with a declining ambient SO_2 concentration, the presence of ambient NH_3 has a progressively larger effect in neutralizing acidity from SO_2 deposited onto external wet surfaces of vegetation.

In summary, the deposition of sulfur remains one of the key components of the acidifying budget, but over the UK it contributes less than half of total acidifying input. The trends with time have revealed evidence of non-linearity which had important policy implications. The cause of the non-linearity appears to be a combination of oxidant limited conversion of SO_2 to $SO_4{}^{2-}$ and a change in deposition velocity with time.

3.2 How Good are the Annual Sulfur Deposition Values?

The combination of modelling, measurements and the number of components in the sulfur deposition budget makes estimates of uncertainty in the annual deposition a complex task. One of the most helpful validation checks on annual deposition is provided by hydrochemical budgets in catchments. Such measurements for conserved species (such as $SO_4{}^{2-}$ and Cl^-) in 'water tight' catchments, in which there are no significant changes in internal storage, require only the outflow volume and composition to deduce inputs. The

Table 2   *Sulfur deposition budgets for Beacon Hill Catchment 1984-88 based on UK deposition maps, modelled by Fowler and Smith (Pers. Comm.), and on measurements made on site (Black and Greenwood, Pers. Comm.)*

UK deposition maps 40 km^2 average S deposition, kg S ha^{-1} y^{-1}	Catchment model 0.622 km^2 S deposition kg S ha^{-1} y^{-1}	Catchment measurement 0.662 km S deposition kg S ha^{-1} y^{-1}
Wet	10.5	10.5	13.2
Cloud	0.2		
Dry	19.5	26.4	17.9*
TOTAL	30.2	36.9	31.2

* Estimated from output - input 1984-88.

exercise is assisted at catchments with on-site wet deposition and cloud chemistry measurements, such as the Plynlimon catchments in central Wales and Beacon Hill in Leicestershire, to permit site specific modelling of deposition inputs. These two catchments are particularly helpful since in the Welsh uplands annual inputs are dominated by wet deposition, whereas in Leicestershire dry deposition is the substantially larger component, and those two catchments test the deposition inputs at opposing ends of the spectrum of partitioning of the inputs into wet and dry deposition. Table 2 summarizes the comparisons between measured sulfur outflow from Beacon Hill catchment for four years and the site specific inputs calculated using the same procedures used to national mapping. For the Welsh catchments, the agreement between measured and modelled sulfur deposition is typically $\pm$ 10 to 15%, for catchments of 10-20 ha[5]. In the case of other ions, the Na^+ budget shows good agreement for the Welsh catchments and this with the sulfur budgets in these uplands provides strong support for the simple model of wet deposition enhancement by seeder-feeder surveying.

4 Oxidized Nitrogen

The annual mean concentrations of NO_2 in rural UK are now substantially larger than those of SO_2 in most areas. The dominance of ground level sources and the much longer atmospheric residence time of NO_2 (especially in winter) than SO_2 are the main cause, since emissions of oxidized nitrogen are smaller than those of sulfur dioxide. However, it is clear that NO_2 is now a major component of the pollution climate of the country. The deposition rates, as mentioned above are smaller than those of SO_2 typically by between a factor of 2 and 3, largely because NO_2 is not deposited onto external surfaces of vegetation and is not very soluble[6,7]. In fact NO_2 is only absorbed at significant rates by stomata and, as a consequence rates of uptake by vegetation are strongly diurnal and seasonal. More than 80% of the NO_2 deposited in the UK

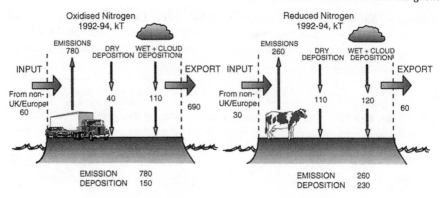

Figure 6   *The UK nitrogen budget for 1992-94*

each year is deposited during the six summer months (April - September). As a consequence of both faster dry deposition and greater oxidation to HNO_3 by gas phase processes in summer, the atmospheric lifetime for NO_2 is substantially shorter in summer (2 to 4 days) than winter (7 to 14 days). The dry deposition model simulates stomatal uptake for the major land uses (arable crops, grassland, moorland, forest and urban areas) and with the wet deposition provides a budget for oxidized nitrogen for the country (Figure 6).

The annual atmospheric inputs amount to 142 kT N and represent a mere 17% of emissions (for 1993), with wet deposition contributing 73% of the input. The remaining oxides of nitrogen are advected out of the country by the wind largely as NO_2 but including particulate nitrate, gaseous HNO_3 and peroxyacetyl nitrate.

Substantial problems of uncertainty in the oxidized nitrogen budget are present as a consequence of large spatial heterogeneity in NO_2 sources and weaknesses in the diffusion tube methodology for rural NO_2 measurement, which is not NO_2 specific and has a tendency to over-estimate actual NO_2 concentration[8]. While these problems lead to uncertainty at individual sites and to some over-estimation in NO_2 at sites with appreciable NO concentrations, it is unlikely that the main conclusions of the atmospheric budget for the UK are significantly wrong. The deposition of oxidized N within the UK is such a small proportion of the emissions that even a factor 2 increase would not change the conclusion. In practice it is likely that the dry deposition has been over-estimated so that the exported fractions may be an even larger fraction of the total.

5 Reduced Nitrogen

The emissions of NH_3 in the UK have proved difficult to estimate precisely due to very large spatial and temporal variability in the conditions which regulate NH_3 release from animal wastes and also from cropland. The current estimate of UK emissions is 263 kT NH_3-N, largely from livestock farming and dairy cows and beef cattle in particular[9]. The inputs of NH_4^+ by wet deposition, like those for SO_4^{2-} and NO_3^- are provided by the precipitation chemistry network and interpolation and modelling of orographic processes to yield a map of NH_4^+ deposition for the country.

The annual input of wet deposited NH_4^+ is 113 kT NH_4-N for 1993. The dry deposition has been much more difficult to measure and model. First there is no reliable NH_3 concentration field from measurements because the available data provided by diffusion tube methods are known to provide large over-estimates in the low concentration areas (< 1 μg NH_3 m^{-3}), which constitute the majority of the UK. The other problem is that NH_3 emissions are extremely heterogeneous. Lastly, NH_3 is a very reactive and soluble gas, whose atmospheric lifetime has been shown to be generally in the range 0.5 to 4 hours, so that mean transport distances are very short (5 to 200 km). The deposition process is also complicated by processes within vegetation which lead to emission from as well as deposition of the NH_3 to plant canopies.

In practice, it is necessary to use a model for land-atmospheric exchange of NH_3 which includes both emission and deposition with the use of a compensation point approach (Figure 5). At ambient concentrations below the compensation point emission fluxes of NH_3 occur and vice versa. To calculate the net exchange of NH_3 over natural and cropland surfaces, a model of emission, transformation, transport and deposition is used to calculate the concentration fields using parameterization from direct field measurements[8]. The land-use dependent dry deposition is then calculated from the concentration field. A new network of NH_3 measurements using simple denuders has recently been established to provide a concentration field directly and validate the modelled concentration patterns.

Using the methods described above, the dry deposition for NH_3 for the UK has been estimated to be 62 kT NH_3-N annually.

From the atmospheric budget of reduced nitrogen over the UK (Figure 7) the total deposited NH_x-nitrogen represents 175 kT-N or 66% of emissions. The comparison against oxidized nitrogen is striking. When expressed in terms of nitrogen, emissions of NH_3-N are smaller than those of NO_2-N and represent only 23% of the total (NO_x-N + NH_3-N) emissions. However reduced nitrogen deposition represents 55% of the total deposited nitrogen. The bulk of dry NH_3 deposition is to semi-natural vegetation (hedgerows, moorland and forest) which in general is the component of the landscape most sensitive to nitrogen inputs.

Table 3 contrasts the emissions lifetimes and contribution to deposited nitrogen of reduced and oxidized nitrogen in the UK. The fact that only a small fraction of NH_3 emitted within the UK is exported leads to an important policy conclusion, that effects of reduced nitrogen deposition in the UK are essentially a UK problem and that the contribution of the UK NH_4-N to long range transport of nitrogen is a small part of the total.

One small but significant argument that needs to be considered is the degree of consistency in the budgets of the three major components of the acidifying

Table 3   *Emissions lifetime and contribution of reduced and oxidised nitrogen to total nitrogen deposition in the UK*

	NO_y-N	NH_x-N
UK Emission (1994)	720 kT N	260 kT N
Lifetime	2-10 days	0.5-5 hours
% UK N deposition	45	55

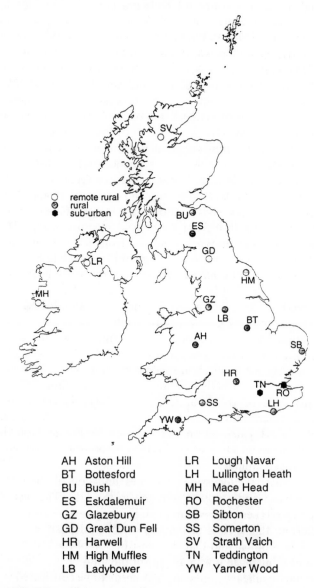

AH	Aston Hill	LR	Lough Navar
BT	Bottesford	LH	Lullington Heath
BU	Bush	MH	Mace Head
ES	Eskdalemuir	RO	Rochester
GZ	Glazebury	SB	Sibton
GD	Great Dun Fell	SS	Somerton
HR	Harwell	SV	Strath Vaich
HM	High Muffles	TN	Teddington
LB	Ladybower	YW	Yarner Wood

Figure 7   *The UK rural ozone network*

compounds. In the case of sulfur, we are fairly confident about the patterns of deposition, the partitioning wet and dry and the total. However, the NH_3 budget has been more problematic, and emission has consistently been difficult to quantify precisely. The UK estimate of 263 kT NH_3-N was concluded after careful consideration of differing scientific opinions, with some arguing for substantially larger values. The atmospheric sulfur budget suggests that about 1000 kT-S is advected out of the country, mainly to the North Sea. If it is assumed that a half of the sulfur is exported as $(NH_4)_2SO_4$, then the export of NH_4^+-N would need to be substantially larger than the 114 kT-N suggested by the reduced nitrogen budget, by as much as a factor of 5. Aerosols sampled in eastern England and Scotland tend to show NH_4^+ and SO_4^{2-} in aerosols in roughly equivalent quantities (expressed as S and N) and with known rates of oxidation of SO_2 to SO_4^{2-} it would be expected that a substantial proportion of the sulfur leaving the UK is as SO_4^{2-}. It seems likely therefore that the 114 kT NH_4-N exported from the UK is an under-estimate, and that as a consequence the UK emissions need to be substantially larger than 263 kT NH_3-N to balance the budget.

Finally, considering the total deposition of nitrogen; the 320 kT of nitrogen deposited in the UK are equivalent to 15 kg N ha^{-1} throughout the country with a range from 5 to 100 kg N ha^{-1}, and the average input exceeds the amounts deposited prior to the industrial revolution by approximately an order of magnitude. The effects of the deposited nitrogen on ecosystem productivity, the release of N_2O and NO are important but the sequestration of atmospheric CO_2 is probably the most important impact and on a regional scale in the boreal forests of the mid-latitude Northern Hemisphere, may be an important contributor to the missing terrestrial sink for atmospheric CO_2.

6 Photochemical Oxidants

The emissions of NO and NO_2 and volatile organic compounds by motor vehicles and industry provide the necessary reactants for the photochemical production of ozone and related photochemical oxidants (including peroxy-acetyl nitrate, HONO and H_2O_2). These oxidants are produced primarily during the spring and summer months. The ambient concentrations of ozone are monitored in rural UK by a network of 18 sites (Figure 8) stretching from Yarner Wood in Devon to Straith Vaich in Sutherland. Unlike SO_2 and NO_2, ozone is strictly a secondary pollutant with no direct emissions at the surface. The periods of intense photochemical activity enhance the natural background concentration within the troposphere, although in the mid-latitude Northern Hemisphere, there is evidence that anthropogenic activity has increased the background from 10 to 12 ppb at the turn of the century to 20 to 25 ppb during the 1980s. The warm sunny weather which characterizes summer anticyclone conditions, is ideal for ozone formation with cloud-free slowly moving air over source regions coupled with high temperatures[9]. In these conditions, the peak concentrations of ozone may reach between 100 ppb and

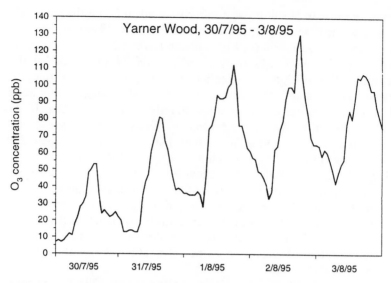

Figure 8   *An ozone episode in southern England in summer 1995*

150 ppb. Figure 8 shows a typical time series of an ozone episode with gradually increasing daily peak values during the 4-day period. In recent years (1985-1994) there is evidence that peak concentrations of ozone have not achieved values as large as those during the previous decade (1974-1984). The data are summarized in Figure 9. These show that the more recent peak ozone concentrations in the UK are 20 ppb to 30 ppb smaller than they were a decade earlier. The important reduction in the exposure to peak values is particularly valuable as it is primarily the peak values which are associated with effects on crops, natural vegetation and human health effects.

A very important part of the analysis of ozone exposure is the mapping of indices of exceedance of thresholds for effects on crops, forests and human health. These concentrations have been extrapolated from a relatively small network of monitoring sites mainly because during the summer daytime conditions the surface ozone concentrations are representative of the bulk of the boundary layer, and in these conditions the surface concentrations vary slowly across the country. The temporal extent of ozone exceedance has also been shown to vary strongly with altitude (Figure 10). Using these properties of the data, maps of exceedance of thresholds for effects have been constructed. Using as an example the AOT_{40} (Accumulated exposure over 40 ppb) maps for crops, the map (Figure 11) shows that a large part of the UK experiences an AOT_{40} in excess of 3000 hours during the average May to July. As this value (3000 ppb hours) is the threshold for effects on arable crops the map shows that the potential for direct effects of ozone is present throughout these regions in the average year, if based on 1993 to 1994 data. Thus a significant crop loss due to ozone may be incurred each year. The extent of AOT_{40} exceedance increases from north-west to south-east across the UK and

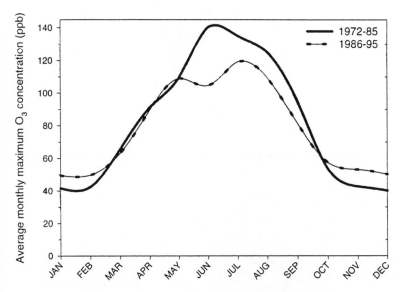

Figure 9   *Peaks of ozone concentration in 1972-85 and 1986-95*

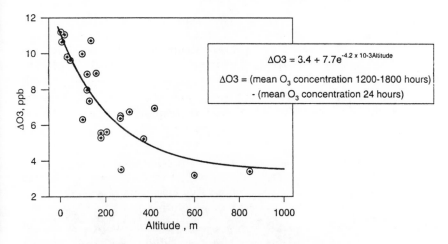

$$\Delta O3 = 3.4 + 7.7e^{-4.2 \times 10^{-3}\text{Altitude}}$$

$\Delta O3 =$ (mean O_3 concentration 1200-1800 hours) - (mean O_3 concentration 24 hours)

Figure 10   *The relationship between ozone concentration and altitude*

extends into Continental Europe with very large exceedances in Southern France, Germany, Switzerland and northern Italy.

The high resolution of the ozone maps results from the simplicity of the meteorological controls over the atmospheric mixing processes in the boundary layer. In practice the heterogeneity of the topography, land use and source distribution for gases which effectively titrate ozone introduce considerable additional uncertainty in these maps. For the broad-scale assessment of environmental effects, the current network requires additional sites to underpin regional spatial patterns, especially in the East Midlands and East

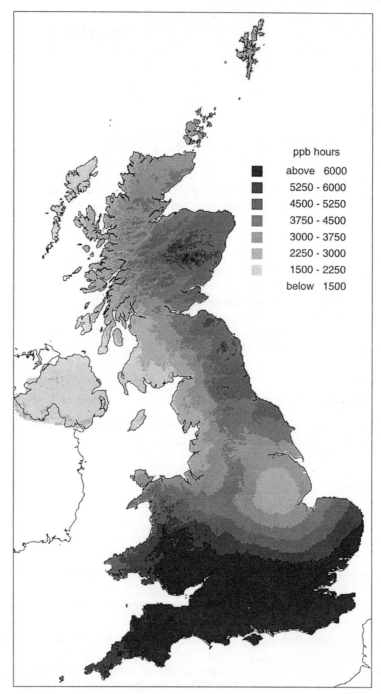

Figure 11   *The UK AOT$_{40}$ ozone exposure map for wheat*

Anglia, the highlands of Scotland and the Welsh uplands. However, the more complex issues of characterizing the exposure of people and materials in urban areas offers a greater challenge and is a necessary step in the next phase of assessment of effects of ozone in the UK.

7 Issues

The preceding pages have briefly highlighted the broad issues of pollutants contributing to acidic deposition, eutrophication effects and photochemical oxidants. These issues and pollutants constitute the major current issues of short-lived pollutants. A full consideration of pollutants emitted in the UK, their sources, date and effects, should extend also to the very large emissions of radiatively active (greenhouse) gases CO_2, CH_4 and N_2O and to the smaller emissions of radioactive releases to the atmosphere. The list should also include heavy metals, persistent organic compounds (POPs) and aerosols in general. These fall outside the remit of this brief overview of the major air pollution regional problems in the UK. Overall sulfur concentrations and depositions are declining and the emphasis on control measures has switched towards the more challenging problems of oxidized and reduced nitrogen, for which a considerable improvement in our understanding of basic processes will be necessary to underpin the development of successful policies.

8 Acknowledgements

The authors acknowledge the support of the Department of the Environment and European Commission for these studies.

References

1. RGAR, Acid Deposition in the United Kingdom 1986-1995. Fourth Report of the UK Review Group on Acid Rain, Department of the Environment, London, 1997.
2. D. Fowler, C. Fletchard, R.L. Storeton-West, M.A. Sutton, K.J. Hargreaves, and R.I. Smith. Long term measurements of SO_2 deposition over vegetation and soil and comparison with models. In: *Acid rain research: do we have enough answers?* edited by G.J. Heij & J.W. Erisman, Amsterdam: Elsevier, 1995, p.9
3. J.W. Erisman, M. Ferm, D. Fowler, L. Granat, H. Hultberg, J. Kesselmeier, R. Koble, G. Lovett, F. Moldan, J. Padro, G. Runge, C. Simmons, S.J. Slanina and M. Zapetal, Dry deposition of sulphur dioxide (Working Group Report). In: *Models and methods of quantification of atmospheric input to ecosystems*, edited by G Lövblad, J.W. Erisman & D. Fowler, Copenhagen: Nordic Council of Ministers, 1993, p.13.
4. J. Löveblad, J-W. Erisman and D. Fowler, Models and methods for the quantification of atmospheric input to ecosystems. Report No Nord 1993: 573. Göteborg, Sweden, November 1992, Copenhagen: Nordic Council of Ministers, 1993.
5. B. Reynolds, D. Fowler and S. Thomas. *Sci. Tot. Environ.*, 1996, **188**, 115.

6. D. Fowler, J.H. Duyzer and D.D. Baldocchi, *Proc. Roy. Soc. Edinburgh* 1991, **97B**, 35.
7. J.H. Duyzer and D. Fowler, *Tellus*, 1994, **46B**, 353.
8. M.R. Heal and J.N. Cape. *Atmos. Environ.*, 1997, (In press).
9. M.A. Sutton, C.E.R. Pitcairn and D. Fowler, *Adv. Ecol. Res.*, **24**, 301.
10. R.J. Singles, M.A. Sutton and K.J. Weston, *Atmos. Environ.*, 1997, (In press).
11. PORG, Department of the Environment, 1997 (In press).

Global Air Pollution Problems - Present and Future

R.G. Derwent

Atmospheric Processes Research, Meteorological Office, London Road, Bracknell, Berkshire RG12 2SZ, UK

1 Introduction

We first became aware that human activities could affect the global atmosphere during the late 1940s and early 1950s, at the time of the atmospheric testing of nuclear weapons. Radionuclides were efficiently mixed throughout the troposphere and were sometimes injected into the stratosphere. The resulting radioactive fall-out was observed throughout the northern hemisphere, with a smaller spread into the southern hemisphere by interhemispheric exchange within the stratosphere.[1] Subsequently, further evidence for global air pollution has come from observations of the increasing concentrations of carbon dioxide,[2] the detection of pesticides such as DDT at remote locations from their point of use[3] and the measurement of man-made halocarbons in the southern hemisphere.[4]

In this chapter, a pollutant is defined as a trace gas that, between the point of its emission into the atmosphere and the point of its ultimate removal, causes harm to a target ecosystem. A global air pollutant is therefore an air pollutant where the emission source and target ecosystem being harmed are in different continents or hemispheres. The mechanism of harm may include: radioactive contamination by radioactive fall-out; ecotoxicity by persistent organic pollutants such as pesticides, their residues and environmental degradation products; ultraviolet radiation damage to human health, plants and aquatic ecosystems through stratospheric ozone layer depletion; exposure to increased concentrations of toxic and aggressive pollutants such as ozone; and damage to habitats and ecosystems through climate change.

Their have been extensive reviews of such global air pollution problems.[5] Their identification has in each case required careful long-term observation of atmospheric trace gas concentrations with high precision at baseline monitoring stations. Such measurements require highly sensitive detection systems and precise calibration techniques. Understanding of the underlying mechanisms involved in global air pollution problems has developed through a combination of field investigation, laboratory studies of detailed chemical and physical processes and computer modelling studies to draw the measurements and process studies together into a common framework.

All of these global air pollution problems came as a complete surprise to the respective scientific communities and none were anticipated in advance. Their control has and will continue to require extensive scientific research to identify the causative agents, quantify their emissions, evaluate abatement strategies and implement them in a fair, rational and cost-effective manner. Environmental improvement can only come from effective legislation firmly based on the underlying environmental science. Global air pollution is an extremely wide and complex subject embracing the disciplines of atmospheric chemistry, meteorology, analytical chemistry, environmental physics, ecology, toxicology and international pollution control. In this chapter attention is given to the detection, concentration and trends in the trace gases which drive two of the main global air pollution problems listed above, namely global climate change and stratospheric ozone layer depletion.

1.1 Global Climate Change

The earth's climate system operates as a balance between the energy received from the sun in short wave radiation and the thermal infrared radiation emitted by the earth and its atmosphere to outer space. If the climate system is in equilibrium, then the absorbed solar energy is exactly balanced by the emitted radiation. Any factor that is able to perturb this balance, and thus potentially alter the climate, is called a radiative forcing agent.[6] Of particular concern in this chapter are the changes in radiative forcing which are brought about from the increases in concentrations of certain radiatively-active trace gases in the troposphere and stratosphere. These are the so-called greenhouse gases, which include carbon dioxide, methane, nitrous oxide and the chlorofluorocarbons.

The ultimate energy source for all weather and climate is radiation from the sun and amounts to about 343 W m^{-2}. Averaged globally and annually, about one third, 103 W m^{-2}, of the incoming solar radiation is reflected back to space. Of the remaining 240 W m^{-2}, some is absorbed by the atmosphere, but most is absorbed by the earth's surface. The solar radiation absorbed by the earth-atmosphere system must be balanced at the top of the atmosphere by the emission of 240 W m^{-2} at infrared wavelengths. Some of the outgoing infrared radiation is absorbed by the naturally-occurring radiatively-active trace gases in the atmosphere, principally water vapour and carbon dioxide, and this natural greenhouse effect keeps the earth-atmosphere system about 33°C warmer than it would otherwise be.[6]

The concentrations of certain radiatively-active trace gases are increasing in concentration. These changes in concentration come about because their emissions or removal mechanisms have been or are or will continue to be influenced by human activities. Under these conditions, their atmospheric concentrations are no longer in steady state with their sources and sinks. The steadily growing concentrations of the radiative forcing agents: carbon dioxide, methane and nitrous oxide are of particular concern. In addition, indirect effects on radiative forcing can result from greenhouse gases that are themselves not emitted into the atmosphere but are formed there by chemical reactions. Indirect radiative forcing is thought to be occurring through the formation of tropospheric ozone and the depletion of stratospheric ozone.

Human activity has also led to an increase in the abundance of aerosols in the troposphere, mainly produced by oxidation of sulfur dioxide from the burning of fossil fuels and by biomass burning. Aerosols can cause a direct radiative forcing through increased scattering and absorption of solar radiation. In addition, they can also exert an influence on the size of cloud droplets and on cloud reflectivity. The radiative forcing effects of aerosols tend to be negative and act to cool the earth's surface. Natural factors, such as an increase in aerosols in the stratosphere produced by volcanic activity can lead to negative radiative forcing and to a cooling of the climate system.[7]

Increasing atmospheric concentrations of radiatively-active trace gases can act to perturb the earth's radiation balance by absorbing either the solar radiation emitted by the sun or the terrestrial radiation emitted by the earth's surface-atmosphere system. This perturbation is called radiative forcing and it leads to an enhanced, man-made greenhouse effect. The radiative forcing agents, both natural and human-influenced, activate physical processes that cause the global climate system to move to a new physical state. The response of the climate system to the trace gas radiative forcing may involve either heating or cooling of the surface-atmosphere system, increased or decreased temperatures, moisture, winds, clouds and ice. This climate system response will then induce a number of biological processes that will cause changes in the world's ecosystems, both natural and managed, and produce a response in them.

Emissions due to human activities are substantially increasing the atmospheric concentrations of the greenhouse gases: carbon dioxide, methane, chlorofluorocarbons and nitrous oxide. These concentration increases will enhance the natural greenhouse effect, resulting in an additional warming of the earth's surface. The main greenhouse gas, water vapour, will increase in concentration in response to the global warming and further enhance it. Global mean temperatures are expected to increase by about 1°C by 2025 and by about 3°C before the end of the next century. There will be a concomitant rise in the global mean sea level of about 20 cm by 2030 and 65 cm by the end of the next century.[6]

The aims of scientific research into global climate change are to understand the influence of human activities on the various forcing agents, to build a predictive understanding of the climate response to these forcing agents and to characterise the impacts that will ensue. Currently we have most confidence in the observed distributions and trends in the concentrations of the radiatively-active trace gases and the extent of human influence upon them. These fundamental observations are the building blocks of our concerns over global warming. Some of this information is presented in the sections below. For further background information on global climate change and the greenhouse effect, the reader is referred to the comprehensive, state-of-the-art reports from the Intergovernmental Panel on Climate Change, IPCC.[6-8]

1.2 Stratospheric Ozone Layer Depletion

In the 1960s the industrialised countries began manufacturing increasing quantities of chlorofluorocarbons, CFCs, and other halocarbons for use as refrigerants, foam blowing agents, aerosol propellants, fire extinguishers and solvents.[9] Although these CFC and other halocarbon molecules are heavier than air, turbulence and atmospheric motions on all scales quickly mix them much faster than they can settle according to their molecular masses. Because they are insoluble in water, they are not removed in the rain-making processes in the troposphere. Those that are unreactive in the troposphere can be been transported into the stratosphere.

Once in the stratosphere, the CFC and other halocarbon molecules are broken down by the high energy solar ultraviolet radiation, releasing chlorine- and bromine-containing breakdown products. These chlorine- and bromine-containing breakdown products remain in the stratosphere for timescales of years to decades where they can catalyse the destruction of stratospheric ozone.[10] In the Antarctic stratosphere during wintertime, temperatures as low as -80°C lead to the formation of polar stratospheric clouds. These clouds facilitate chemical reactions which allow the conversion of the chlorine-containing breakdown products into active chlorine compounds which, together with active bromine compounds, start to destroy ozone photochemically during the polar sunrise. During August and September, at the end of the winter and the beginning of the spring in the Southern Hemisphere, sunlight driven chemical processes begin the destruction of the polar ozone and the formation of the ozone hole. By the middle of October, ozone is completely removed in the 14-19 km layer and an 'ozone hole' extends to cover much of the stratosphere above the Antarctic continent. Atmospheric motions fill in the 'ozone hole' steadily during November so that by early December, all traces have usually disappeared.

Most of the chlorine and bromine in the stratosphere is there as a direct result of human activities. Many compounds containing chlorine are released at the earth's surface, but those that are water-soluble cannot reach the Antarctic stratosphere. So the large quantities of chlorine that are released as sodium chloride aerosol by the evaporation of sea spray dissolve in rain and exert no influence on stratospheric chemistry. In contrast, the man-made chlorofluorocarbons, such as CFC-11 and CFC-12, and the other halocarbons, such as carbon tetrachloride and methyl chloroform, are not water-soluble and are not broken down in the troposphere. Currently, 88% of the chlorine entering the stratosphere is man-made in origin, with 12% from natural sources. The main man-made stratospheric chlorine carriers are: CFC-11, 28%;

CFC-12, 23%; carbon tetrachloride, 12%; methyl chloroform, 10%. The main natural chlorine carrier is methyl chloride which accounts for 15% of the chlorine entering the stratosphere currently from all sources, both natural and human-influenced.[10]

The ozone depletion caused by man-made chlorofluorocarbons and other halocarbons is expected to persist until chlorine and bromine levels are reduced by stratospheric destruction. World-wide monitoring has shown that stratospheric ozone has been decreasing for the past two decades, with the overall change since the 1960s currently close to 5% globally. This phenomenon is called mid-latitude ozone depletion and it is clearly distinguished from polar ozone depletion in Antarctic 'ozone holes'. Since the late 1970s, an 'ozone hole' has formed in Antarctica during September and October in which up to 60% of the total ozone is depleted. The large increase in stratospheric concentrations of man-made chlorine and bromine carriers is responsible for the formation of the Antarctic 'ozone hole' and plays a major role in mid-latitude ozone depletion.

In 1987, the recognition that chlorine and bromine had the potential to destroy stratospheric ozone led to an international agreement, The United Nations Montreal Protocol on Substances that Deplete the Ozone Layer. Under this treaty and its various amendments, steps are being taken to reduce and eventually to eliminate the global production of ozone-depleting substances, particularly the chlorofluorocarbons and other halocarbons mentioned above. The atmospheric abundances of the main chlorine and bromine carriers to the stratosphere are expected to reach their maximum concentrations by the turn of the century and then start to decline. All other things being equal, the stratospheric ozone layer should start to recover by the middle of the next century.

The aims of scientific research into stratospheric ozone depletion are to understand the influence of human activities on the various chlorine and bromine carriers to the stratosphere, to build a predictive understanding of the response of the ozone layer to these carriers and to characterise the impacts that ensue. Some of the information on the growth in tropospheric concentrations of the chlorofluorocarbons and halocarbons is presented in the sections below. For further background information on stratospheric ozone layer depletion, the reader is referred to the comprehensive, state-of-the-art reports from the World Meteorological Organization and the United Nations Environment Programme.[10-13]

2 Atmospheric Composition

The composition of the dry unpolluted atmosphere is shown in Table 1 in terms of the by volume concentrations of the eleven most common constituents[14]. The mass of the dry atmosphere can be taken[15] as 5.132×10^{18} kg and the mean atmospheric pressure as 982.4 mb. The mean mass of water vapour is

Table 1   *The composition of sea-level dry unpolluted air by volume.*[14]

Species	Concentration by volume
nitrogen N_2	78.084%
oxygen O_2	20.9476%
argon Ar	0.934%
carbon dioxide CO_2	360 ppm (variable)
neon Ne	18.18 ppm
helium He	5.24 ppm
methane CH_4	1.7 ppm (variable)
krypton Kr	1.14 ppm (variable)
hydrogen H_2	0.5 ppm (variable)
nitrous oxide N_2O	310 ppb (variable)
xenon Xe	87 ppb

1.35×10^{16} kg. From the composition in Table 1, an average molecular mass of the atmosphere is estimated as 28.9644 mol/g.

The most common units for reporting atmospheric trace gas concentrations are as volume mixing ratios. By volume mixing ratio, we mean the ratio of the volume of that component to the total volume of air. This definition is exactly the same as the chemist's mole fraction. For the permanent gases, mixing ratios are given as ratios or as percentages. For the trace constituents, parts in one million parts or more concisely: parts per million, ppm, are frequently used. With increasing analytical precision, more and more species have been detected in the atmosphere at even lower concentrations. The parts per billion, or more strictly parts in 10^9 parts or ppb and parts per trillion: parts in 10^{12} parts or ppt are now the more commonly used concentration units.

3 Global Concentrations and Their Trends for the Main Greenhouse Gases

The current concentrations and concentration trends for the major radiatively-active trace gases or greenhouse gases are collected together in Table 2. Much of this material has been drawn from the reports of the Intergovernmental Panel on Climate Change.[6-8] Current assessments of radiative forcing place the man-made greenhouse gases in the order of importance: carbon dioxide > methane > nitrous oxide > tropospheric ozone. This ordering reflects a combination of the influences of emissions, atmospheric lifetimes and radiative forcing per molecule which are different for each greenhouse gas.

3.1 Carbon Dioxide

Measurements of carbon dioxide began at the South Pole in 1957 and at Mauna Loa, Hawaii in 1958.[2] The atmospheric concentration was then about 315 ppm and the rate of increase about 0.6 ppm per year. The concentrations

Table 2 *Global mean concentrations and current growth rate trends for the major greenhouse gases.*[2]

Species	Global mean concentration in 1992	Current growth rate
tropospheric water H_2O	10-20,000 ppm	
stratospheric water H_2O	2-6 ppm	
carbon dioxide CO_2	356 ppm	1.6 ppm per year
methane CH_4	1714 ppb	8 ppb per year
nitrous oxide N_2O	311 ppb	0.8 ppb per year
stratospheric ozone O_3	200-10,000 ppb	
tropospheric ozone O_3	10-200 ppb	
sulfur hexafluoride SF_6	32 ppt	0.2 ppt per year
carbon tetrafluoride CF_4	70 ppt	1.2 ppt per year
hexafluoroethane C_2F_6	4 ppt	

at Mauna Loa have increased steadily each year at about 0.83 ppm per year in the 1960s, 1.28 ppm per year in the 1970s and about 1.53 ppm per year in the 1980s. During the early 1990s the growth rate dropped to 0.6 ppm per year but recovered markedly during 1993 and 1994 to levels above the long term mean. In 1994 the global mean concentration was about 358 ppm, with a trend of about 1.6 ppm per year.[8]

Prior to 1957, atmospheric carbon dioxide records mainly come from air bubbles trapped in ice cores. Over the last 1000 years, concentrations have fluctuated by about ± 10 ppm around a mean of 280 ppm. Throughout these records there is no evidence that past changes were as rapid as those observed from 1950 onwards. Concentrations started to rise from their pre-industrial level of 278 ppm around 1800 and had already risen by 15 ppm by 1900. Current levels are the highest ever observed.[8]

The major components of the atmospheric carbon dioxide budget during the 10 year period from 1980-1989 have been assembled from emission inventories, atmospheric measurements, modelling, measurement of isotopic ratios and satellite observations. Emissions from fossil fuel combustion and cement manufacture averaged 5.5 Gtonne C per year and from changes in tropical land use, 1.6 Gtonne C per year. Total sources were therefore about 7.1 Gtonne C per year. Storage in the atmosphere amounted to 3.4 Gtonne C per year and uptake by the oceans accounted for a further 2.0 Gtonne C per year. This leaves an additional 1.7 Gtonne C per year sink to be accounted for, presumably in terrestrial ecosystems of the northern hemisphere. Over the 1980-1989 period, therefore, roughly 60% of the fossil fuel emissions of carbon dioxide remained in the atmosphere.[8]

3.2 Methane

Extensive measurements of methane began in 1978 or thereabouts when global mean concentrations were about 1516 ppb, increasing at about 20 ppb per year.[16] More extensive measurements began in 1984 and these have indicated lower growth rates of 13 ppb per year during the 1980s, slowing to 9 ppb per year by 1991. During 1992 and 1993, methane stopped growing in some locations for reasons which have not been quantitatively understood, but which are presumably connected with the anomalous behaviour reported for carbon dioxide and nitrous oxide. In 1994 and 1995, the methane growth rates have reestablished themselves at about 8 ppb per year, bringing the global mean concentration up to 1721 ppb in 1994.[8]

Measurements of methane in air bubbles trapped in ice cores confirm a pre-industrial global concentration of 720-740 ppb, with northern hemispheric values some 35-75 ppb higher than southern hemispheric values. Methane concentrations began increasing from their pre-industrial values in about 1700 and reached 900 ppb by about 1900. Current levels are the highest ever seen for methane in ice core records which go back 160,000 years.[8]

The major natural sources of methane appear to be wetlands, accounting for 115 Tg per year out of 160 Tg per year. Human activities emit 375 Tg per year, of which natural gas, coal mining, cows, rice paddies, biomass burning and landfills are all significant sources. The total methane source strength is currently about 535 Tg per year and the annual increase in the atmospheric burden is about 37 Tg per year. For a total global burden of 4850 Tg, this implies a methane lifetime of about 9 years.

3.3 Nitrous Oxide

The global mean nitrous oxide concentration was 275 ppb in 1951, 299 ppb in 1976 and showed a growth rate of 0.55 ppb per year during the 1976-1982 period. The growth rate increased to about 0.8 ppb per year in 1989 and then slowed somewhat to the current rate of increase of 0.5 ppb per year during the 1990s, covering the period of the anomalous trends in carbon dioxide and methane. The global mean nitrous oxide concentration in 1992 was 311 ppb.

Ice core records indicate pre-industrial nitrous oxide concentrations of 260-285 ppb, some 15% below present day levels. Historically, soils appear to have been the major natural source. Cultivated soils, biomass burning and industrial sources are now the main sources from human activities. The total source strength is 14.7 Tg N per year and lifetime 120 years.[8]

3.4 Alternative or Replacement Halocarbons

A number of alternative halocarbons have been suggested for use as replacements for the ozone-depleting chlorofluorocarbons and other halocarbons. Generally speaking, these alternative halocarbons contain hydrogen and fluorine but no chlorine. They have the potential to be radiatively-active gases should they have long enough lifetimes and their atmospheric releases become appreciable. For most members of this class of halocarbon, measurement data are too few to allow the compilation of global mean concentrations as in Table 2. Some of the halocarbons proposed or in use as alternatives are listed below and each has a radiative forcing potential on a mass emitted basis which is at least one hundred times larger than that of carbon dioxide.

Some of the alternatives are:

HFC-23 CHF_3
HFC-32 CH_2F_2
HFC-41 CH_3F
HFC-43-10mee $C_5H_2F_{10}$
HFC-125 C_2HF_5
HFC-134 CF_2HCF_2H
HFC-134a CH_2FCF_3
HFC-143 CF_2HCH_2F
HFC-143a CH_3CF_3
HFC-152a CH_3CHF_2
HFC-227ea C_3HF_7
HFC-236fa $C_3H_2F_6$
HFC-245ca $C_3H_3F_5$
HFOC-125e CF_3OCHF_2

3.4 Perhalogenated Species

Table 2 contains some fragmentary data for the perhalogenated trace gases: carbon tetrafluoride CF_4, hexafluoroethane or perfluoroethane C_2F_6 and sulfur hexafluoride SF_6. Each of these species strongly absorb infrared radiation and have exceedingly long atmospheric lifetimes. It is likely that the importance of these trace gas species will rise steadily in importance over the next decade as their concentrations build up in the global atmosphere.[8]

4 Global Concentrations and Their Trends for the Ozone-depleting Chlorofluorocarbons and Other Halocarbons

The current tropospheric concentrations and concentration trends for the major ozone-depleting chlorofluorocarbons and other halocarbons are collected together in Table 3. Much of this material has been drawn together from the reports of the World Meteorological Organization and the United Nations Environment Programme.[10-13] Current assessments place the major man-made chlorine carriers in the order of importance: CFC-12 > CFC-11 > CCl4 > methyl chloroform > CFC-113 > HCFC-22.

Recent trends in the tropospheric concentrations of the major chlorine carriers to the stratosphere are important in understanding stratospheric ozone layer depletion. Estimates of budgets and lifetimes are required to assess future impacts, and to predict what levels of emission reductions are needed to stabilise or to reduce these impacts.

4.1 Chlorofluorocarbons

CFC-11 is a completely man-made substance that was not present in the atmosphere prior to its commercial manufacture. By the time the ALE global measurement programme began in 1978 the global concentration had already reached 160 ppt and it was growing at 9-11 ppt per year during the 1970s and 1980s.[16] By 1990, northern hemisphere concentrations stopped growing and started to decrease in response to the United Nations Montreal Protocol. By 1992 the global mean concentration had stabilised at 268 ppt or thereabouts and has subsequently started to fall.

The peak concentration corresponds to a total global burden of 6.2 Tg and can be directly compared with the total integrated release to the atmosphere at that time of 7.2 Tg.[17] The difference between the peak burden and the integrated release corresponds to the integrated amount of CFC-11 which has been lost to sinks. The main sink process is stratospheric photolysis with a lifetime of 50 $\pm$ 5 years.[8]

CFC-12 has shown similar behaviour to that of CFC-11 and this is not surprising since its atmospheric life cycle shares many common features. In 1978 its global concentration had already reached 265 ppt and its concentration was rising at 16-20 ppt per year.[16] By 1992 the global concentration had reached 503 ppt but the growth rate had slowed to 7 ppt per year. It is anticipated that the global CFC-12 concentrations will continue to rise for a number of years after its phase-out under the United Nations Montreal Protocol because of the 'bank' of sales into refrigerators and the like which are still in use.

By 1992 the global burden of CFC-12 amounted to 10.3 Tg, a figure which is close to the total integrated atmospheric release of 10.1-11.2 Tg.[17] It is likely that the close agreement between these figures disguises the small loss due to stratospheric photolysis which fixes the CFC-12 lifetime at 102 years or thereabouts.[8]

CFC-113 is of relatively recent commercial manufacture and use compared with CFC-11 and -12. Its global mean concentration had reached 82 ppt by 1992 with a growth rate that averaged 5-6 ppt per year during the late 1980s. This growth rate slowed markedly during the early 1990s and the global concentration has stabilised at around 84 ppt. Northern hemispheric concentrations have begun to decline in 1994 in response to the global phase-out under the United Nations Montreal Protocol. At its peak, the atmospheric burden of CFC-113 amounted to about 2.7 Tg, a figure which can be compared with the total atmospheric release of 3.3 Tg.[18]

Table 3 *Global concentrations and concentration trends for the major man-made chlorofluorocarbons and other halocarbons.*[8,10]

Species	Global mean concentration in 1992	Current growth rate
Species phased out under the United Nations Montreal Protocol and its amendments		
CFC-11 CCl₃F	268 ppt	decreasing
CFC-12 CCl₂F₂	503 ppt	7 ppt per year
CFC-113 CCl₂FCClF₂	82 ppt	decreasing
CFC-114 CClF₂CClF₂	20 ppt	
CFC-115 CF₃CClF₂	<10 ppt	
carbon tetrachloride CCl₄	132 ppt	decreasing
methyl chloroform CH₃CCl₃	135 ppt	decreasing
halon-1211 CBrClF₂	7 ppt	0.15 ppt per year
halon-1301 CBrF₃	3 ppt	0.2 ppt per year
halon-2402 CBrF₂CBrF₂	0.7 ppt	
Species controlled by the United Nations Montreal Protocol and its amendments		
HCFC-22 CHClF₂	100 ppt	5 ppt per year
HCFC-123 CF₃CHCl₂		
HCFC-124 CF₃CHClF		
HCFC-141b CH₃CFCl₂	2 ppt	1 ppt per year
HCFC-142b CH₃CFCl₂	6 ppt	1 ppt per year
HCFC-225ca C₃HF₅Cl₂		
HCFC-225cb C₃HF₅Cl₂		

4.2 Other Halocarbons

Carbon tetrachloride is one of the oldest industrial halocarbons and has been used as an intermediate in the manufacturing process for CFC-11 and CFC-12 and as a general purpose solvent. When the ALE network was established its global mean concentration was reported to be about 112 ppt with a global trend of about 1.5 ppt per year.[16] The global mean concentration reached its peak of 132 ppt in 1990-2 and has been declining ever since in response to the phase-out in CFC production under the United Nations Montreal Protocol. The global burden at its peak amounted to about 3.4 Tg.[8]

Methyl chloroform is a major stratospheric chlorine carrier and has been widely used in industry as a degreasing solvent prior to its global phase-out under the United Nations Montreal protocol and its amendments. It differs from all of the other phased-out species detailed above because its has an efficient tropospheric sink process: oxidation by hydroxyl radicals. The methyl chloroform lifetime due to hydroxyl radical destruction and other sink processes is currently estimated as 4.8 ± 0.2 years.[19] Despite this efficient removal process, man-made emissions have grown to the extent that methyl chloroform concentrations have steadily increased since global measurements began in 1978.

At their peak during 1991-2, global mean concentrations of methyl chloroform reached 153 ppt, corresponding to a global burden of 3.3 Tg. The total integrated atmospheric release up until that time had been 11.6 Tg,[20] with the difference from the global burden being accounted for by destruction by reaction with tropospheric hydroxyl radicals.

The most important natural chlorine carrier to the stratosphere is methyl chloride. Its global mean concentration is about 600 ppt and it currently (mid-1990s) carries about 15% of the chlorine to the stratosphere from all sources both natural and man-made.[10] Measurements made over the period from the late 1970s to the mid 1980s have shown no long-term trends. Methyl chloride is released from the oceans as a natural source and from biomass burning as a human-influenced source. The main removal process for methyl chloride is destruction by reaction with tropospheric hydroxyl radicals for which a lifetime of 18 months is estimated.[8]

Methyl bromide is the main carrier of bromine into the stratosphere along with the halons. Its tropospheric concentrations lie in the range 5-30 ppt, though measurements made since 1985 are generally in the range 8-15 ppt.[10] The main removal process for methyl bromide is destruction by reaction with tropospheric hydroxyl radicals, for which a lifetime of 14 months results.[8] The main sources of methyl bromide appear to be the oceans, biomass burning, agricultural fumigation, and petrol-engined vehicle exhausts.[10]

Acknowledgements

The support of the Global Atmosphere Division of the Department of the Environment under the research programme contract EPG 1/1/23 is gratefully acknowledged. This study would not have been possible without the encouragement of Dr Peter Simmonds, Dr Graham Nickless and Dr Simon O'Doherty of the School of Chemistry, University of Bristol.

References

1. L. Machta, 'Advances in Geophysics', Academic Press, London, 1959, Vol. 6, p. 273-286.
2. C.D. Keeling, R.B. Bacastow, A.F. Carter, S.C. Piper, T.P. Whorf, M.Heimann, W.G. Mook, H. Roeloffzen, 'Geophysical Monograph', AGU, Washington DC, 1989, Vol. 55, p. 165-236.
3. J.M. Pacyna and M. Oehme, *Atmos. Environ.*, 1988, **22**, 243-257.
4. J.E. Lovelock, R.J. Maggs and R.J. Wade, *Nature*, 1973, 241, 194-196.
5. C.N. Hewitt and W.T. Sturges, 'Global Atmospheric Chemical Change', Elsevier Applied Science, London, 1993.
6. J.T. Houghton, G.J. Jenkins and J.J. Ephraums, 'Climate Change. The IPCC Scientific Assessment', Cambridge University Press, Cambridge, 1990.
7. J.T. Houghton,L.G. Meira Filho, J. Bruce, H. Lee, B.A. Callander, E. Haites, N. Harris and K. Maskell, 'Climate Change 1994', Cambridge University Press, Cambridge, 1995.
8. J.T. Houghton, L.G. Meira Filho, B.A. Callander, N. Harris, A. Kattenburg and

K. Maskell, 'Climate Change 1995. The Science of Climate Change', Cambridge Univesity Press, Cambridge, 1996.

9. R.L. McCarthy, F.A. Bower and J.P. Jesson, *Atmos. Environ.*, 1977, **11**, 491-497.

10. D.L. Albritton, R.T. Watson and P.J. Aucamp, 'Scientific Assessment of Ozone Depletion: 1994', World Meteorological Organization, Global Ozone Research and Monitoring Project - Report No. 37, Geneva, 1995.

11. 'The Stratosphere 1981 Theory and Measurements', World Meteorological Organization, Global Ozone Research and Monitoring Project - Report No. 11, Geneva, 1981.

12. 'Scientific Assessment of Stratospheric Ozone: 1989', World Meteorological Organization, Global Ozone Research and Monitoring Project - Report No. 20, Geneva, 1989.

13. 'Scientific Assessment of Stratospheric Ozone: 1991', World Meteorological Organization, Global Ozone Research and Monitoring Project - Report No. 25, Geneva, 1989.

14. 'U.S. Standard Atmosphere, 1976', National Oceanic and Atmospheric Administration, Washington DC, 1976.

15. K. E. Trenberth and C.J. Guillemot, *J. Geophys. Res.*, 1994, **99**, 23079-23088.

16. F.S. Rowland and I.S.A. Isaksen, 'The Changing Atmosphere', Wiley-Interscience, Chichester, 1988.

17. D.M. Cunnold, P.J. Fraser, R.F. Weiss, R.G. Prinn, P.G. Simmonds, B.R. Miller, F.N. Alyea and A.J. Crawford, *J. Geophys. Res.*, 1994, **99**, 1107-1126.

18. P.Fraser, D. Cunnold, F. Alyea, R. Weiss, R. Prinn, P. Simmonds, B. Miller and R. Langenfelds, *J. Geophys. Res.*, 1996, **101**, 12585-12599.

19. R. Prinn, R. Weiss, B. Miller, J. Huang, F. Alyea, D. Cunnold, P. Fraser, D. Hartley and P. Simmonds, *Science*, 1995, **269**, 187-192.

20. P.M. Midgley and A. McCulloch, *Atmos. Environ.*, 1995, **29**, 1601-1608.

The Health Effects of Air Pollution in the United Kingdom

J.G. Ayres

Chest Research Institute, Birmingham Heartlands Hospital, Birmingham
B9 5SS, UK

1 Introduction

Air pollution has been very high on the public agenda of late, fuelled by the media who have frequently overstated the true health effects of air pollution. There is no doubt that air pollution at current levels can cause health effects, but suggestions that tens of thousands of people a year are dying purely from air pollution are quite unfounded. These stories have largely derived from the extrapolation of experimental results from the United States to the United Kingdom, which, in view of the wide disparity both qualitatively and quantitatively of air pollutants in those two countries, is inherently unwise.

At one stage the UK led the World in air pollution research, consequent upon the London fog incident of 1952 which caused 4,000 excess deaths over and above predicted in a week[1]. This led to the passage of the Clean Air Act in 1956 and by the early 1960s the air was becoming visibly cleaner. This led, in turn, to a false sense of security and a belief in Government that air pollution was a thing of the past as far as health was concerned and consequently air pollution research effectively ceased in the UK.

However, during the 1970s and 1980s it became apparent that air quality was poor but that the sources of air pollution were clearly different. In the 1950s the sources were largely domestic coal fire burning and industry, whereas nowadays the main sources are motor vehicles.

We are now beginning to appreciate the contribution air pollution is making to long-term morbidity and exactly what proportion that is when compared to other factors. We are slowly beginning to work out the contribution of air pollution to exacerbations of asthma, exacerbations of chronic obstructive pulmonary disease, perhaps the prevalence of asthma and chronic obstructive pulmonary disease, allergic rhinitis (hay fever) and the impact on patients with pre-existing coronary heart disease and cerebro-vascular disease.

2 Methods of Determining Health Effects

The major diseases which are associated with air pollution are those of the respiratory tract and, perhaps surprisingly, coronary artery disease and stroke. However, the vast bulk of information on health effects addresses lung disease which is the major area for concern. It is of course important to realise that there are many other factors that are responsible for respiratory ill health, notably cigarette smoke, allergen exposure and viral infections, and for heart disease cigarettes, diet and exercise. Consequently, when one appreciates that the size of the health effects of air pollution on individuals is relatively small (and very small when compared for instance to the effects of cigarette smoke on lung disease) then it becomes increasingly important when trying to assess the size of health effects from air pollution to ensure that all other contributory factors have been adequately assessed. Another factor which on the face of it would seem to be important is the time spent breathing indoor as opposed to outdoor air. The vast majority, probably 80-90%, of our time is spent indoors, but penetration of air pollution from out to in is in fact significant for many pollutants although there are clear, very important indoor sources which have be considered, notably nitrogen dioxide from gas burning stoves and particles from environmental tobacco smoke.

There are a number of ways of determining health effects in man. Most of the information we have comes from epidemiological studies. However, there is a considerable body of evidence obtained from challenging patients or normal subjects with individual pollutants or combinations of pollutants in defined exposures for specified times. Animal work can help define mechanistic aspects of the health effects of air pollutants while more recently in-vitro work of isolated cell cultures or lavage fluid obtained from the nose or the lung have furthered our knowledge in this area. The final way in which health effects can be assessed is by computer modelling, particularly when it comes to personal exposures to specified pollutants which can be extremely difficult to measure directly.

2.1 Epidemiological Studies

The bulk of epidemiological evidence comes from studies of hospital admissions or mortality, data which is routinely collected and which therefore suffers from the problem of not being individualised i.e. without personal data on individual risk factors for disease. This would increase the sensitivity of such studies when trying to define more clearly those demographic details which might be important in determining health effects. Cross-sectional studies take populations of differing pollution exposure and measure the prevalence of disease within those groups. This can lead to a certain amount of over-interpretation, particularly when only two areas (for instance two towns, one of low and one of high air pollution) are studied. The theoretical problem is that although the one high pollution town may show, for instance, a high prevalence of respiratory disease and the reverse for the selected low pollution town, these may each represent the upper and lower ends of the range of disease prevalences from a range of towns of equivalent pollution experience. Nevertheless these studies have proved useful in assessing such effects because a consistent, repeated finding from a series of 'two town' studies would support any relevant measured association. At a more individual level, panel or event studies involve close assessment of panels of patients who record day to day symptoms and measures of lung function which are then analysed by multiple regression analysis against day to day variations in air pollution. Simpler statistical analysis should not be used in these studies where the effects are small and may not even be noticed by the individuals involved.

2.2 Challenge Studies

These studies, while being intuitively attractive, suffer from the difficulty of producing adequate challenge systems. Challenge rooms are expensive while closed challenge systems such as head domes or mouth-piece delivery systems can only be used for short periods of time. Although multiple exposures and co-exposures to other factors such as exercise can be incorporated in these studies, this does tend to lead to a reductionist approach when interpreting results, perhaps taking attention away from the likelihood that it is the pollution mix which may be more important than the individual contributors. Nevertheless, a considerable amount of data has been obtained which have clarified what health effect each pollutant is able to cause and at what dose threshold such effects occur.

2.3 Animal Studies

Although animal studies have been of help in determining the effects of air pollution there is always a difficulty in extrapolating health effects in animals to humans. In many studies normal rather than compromised animals have been used and consequently potentially useful lines of enquiry may have been ignored when high dose exposures in normal animals did not appear to have an effect. With the advent of more sophisticated in-vitro techniques it may be that animal studies in the future will be less informative.

2.4 In-Vitro Studies

It is now possible to keep human tissue alive in vitro as cell mono-layers which have enabled studies of exposure in the long term to be studied with respect to the way that such cells can be damaged. It has to be appreciated that these are isolated cell cultures without the usual repair mechanisms of the intact human, but nevertheless very interesting data are beginning to be derived from such studies.

2.5 Computer Modelling

Because precise personal exposures are difficult to define due to changes in ventilation rates, indoor/outdoor dwell times and co-exposures (eg environmental tobacco smoke) computer modelling has been developed to estimate personal exposures. Such modelling principles have also been applied to estimations of ambient pollutant levels by knowledge of traffic densities and flows and topographical details of the areas in which health effect studies are conducted.

3 Air Pollution and Respiratory Disease

Diseases of the lung are very common. It is estimated that nearly 20% of the NHS budget is spent on patients with diseases of the respiratory tract and that 1 in 5 consultations with a general practitioner is due to diseases ranging from rhinitis to bronchitis to pneumonia[2]. The bulk of respiratory disease in general practice emanates from the upper respiratory tract (nasal disease, sore throats, tonsillitis) but there is a very significant component to a general practitioners work load from asthma and chronic obstructive airways disease (COAD) a group of diseases which incorporates chronic bronchitis and emphysema. By contrast, in hospital practice the bulk of the work load comprises asthma and COAD, along with contributions from lung cancer and pneumonia.

3.1 Respiratory Disease

Asthma affects around 6% of the UK population and perhaps as many as 20% of primary school children at any one time[3]. COAD is largely due to cigarette smoking although other occupational environmental factors may have played a part. In England and Wales over 25,000 deaths occur from this condition annually, although most of these deaths are in patients over the age of 60[4]. The reason that asthma has been employed in studies of air pollution induced health effects is because the intrinsic problem in patients with asthma is that they have irritable airways due to a persistent inflammatory reaction in their bronchial tubes. This bronchial reactivity can then be acted on by increases in air pollution allowing the asthmatic airway to contract and produce symptoms. The asthmatic patient thus acts as a sort of biological litmus paper.

The way in which an individual responds to an air pollutant will depend on a number of factors. The dose (calculated by the level of exposure and duration of exposure) is clearly important as is exposure to another pollutant at the same time; when considering health effects it is important to realise that one does not inhale single pollutants in isolation when breathing ambient air. There are other factors which are known to cause the airways to react, particularly asthmatic airways, such as viral infection, breathing cold air, exercise, and inhaling cigarette smoke (either actively or passively). Host factors are also important, particularly the degree of bronchial reactivity, whether the patient is atopic (i.e. can generate allergic reactions) and the degree of airflow obstruction. In a condition such as chronic airway narrowing (i.e. COPD) one patient may have more airflow obstruction than the next. Equally, in a patchy condition such as emphysema where some parts of the lung are relatively normal whilst other parts have such narrowed airways that a pollutant could penetrate only poorly, the relatively healthy parts of the lung are more likely to receive a larger dose of an inhaled pollutant, as the airways are open and more able to be exposed.

3.2 Asthma

There is a considerable body of evidence which suggests that day to day changes in air pollution can cause minor changes in symptoms and lung function in both children and adults with asthma, but there is very limited evidence to suggest that individuals exposed to air pollution in the long term are more likely to become asthmatic than had they not been so exposed[5].

3.2.1 Epidemiology.   The main evidence for health effects come from a series of panel studies largely conducted in North America in children. These suggest that on a day-to-day basis peak flow (a simple measure of lung function) and symptoms vary by the order of 3% in asthmatic or susceptible individuals[6] during summer ozone episodes. Some changes in lung function have also been seen although to a lesser degree in normal children. In adults the picture is somewhat less clear, probably due to the effects of other variables having caused (or being a continuing cause of) lung problems. In the UK two panel studies, one in Manchester[7] and one in Birmingham[8], have shown similar changes in adults, although the Manchester study did not take into considera-tion the effects of particles. The Birmingham study showed that the most important predictor of an effect was aerosol strong acid, notably as sulfuric acid (contributed to by sulfur dioxide) and nitric acid (from nitrogen dioxide).

Two air pollution events in 1991 in London[9] and in 1993 in Birmingham[10] were important in that exposures over a period of 2-3 days to very high levels of pollution were seen. In the London event nitrogen dioxide levels rose to 423 ppb (hourly average) and particles rose to in excess of 200 $\mu g/m^3$. Despite this, there was no increase in hospital admissions for asthma, nor any increase in attendances to general practitioners. There were, however, increases in all-cause mortality (by 10%) and hospital admissions for COPD (by 14%). In the Birmingham event where NO_2 levels did not climb quite so high but where sulfur dioxide levels exceeded 200 ppb, patients with severe asthma showed a fall in their lung function and an increase in symptoms despite increasing treatment substantially, whereas patients with mild asthma showed no signifi-cant changes in lung function, symptoms or treatment, suggesting that patients with more severe disease are the most susceptible to changes in air pollution. The duration of these effects is thought to be short when considering day to day effects, probably lasting a matter of hours. However, some winter episodes have, on occasion been associated with persistent decrements in lung function lasting two weeks or so.

There have been two studies of hospital admissions in the UK - one from London[11] and one from Birmingham[12]. The London study showed a significant effect of ozone with a 6% increase in hospital admissions for a 29 ppb rise in ozone. The Birmingham study showed a 0.5% increase in respiratory admissions for a 10 $\mu g/m^3$ rise in either black smoke or sulfur dioxide.

When attempting to dissect out from the epidemiological evidence which of the pollutants may be the most important with respect to health effects, the picture is far from clear. For children, during summer episodes, ozone and particulates (particularly aerosol strong acid) are the most important, while the Birmingham UK study of adults with asthma[8] identified aerosol strong acid, and to a lesser extent nitrogen dioxide, as being important. Health effects from some panel and event studies during the winter seem to be more associated with particles, while in other summer studies, nitrogen dioxide sometimes (but not always) emerges as a possible factor. As far as the hospital admission and mortality studies are concerned, particles (now usually measured as that fraction less than 10 μm in diameter, PM_{10}) emerge as the most important, although with variable contributions from gaseous pollutants. However, in the UK studies, while particles (either as PM_{10} or black smoke) are associated with hospital admissions, the mortality associations, so consistently found in the USA[13], are much less strong. A recent study of data from the city of Amsterdam[14] has revealed no significant association of PM_{10} with mortality but a stronger association with black smoke. There is good reason to believe that black smoke is a better indicator than PM_{10} of ultrafine particles (<2.5 μm) which suggests that a different metric for assessment of particle exposure may be needed. It is possible that particle numbers, rather than mass, may be a more important driver of health effects. If one considers two cubic metres of air, each containing 10 $\mu g/m^3$ of particle by mass, one might contain 5 million particles and another 50,000. The surface area of the sample with more, smaller particles will be considerably the greater and will have a greater chance of affecting a greater area of lung surface.

The above studies confirm that air pollution does have the ability to exacerbate asthma but there have been no UK studies which have addressed the possibility that long-term exposures can cause asthma de *novo* in normal subjects. The only evidence for this comes from the study of Seventh Day Adventists in the United States[15] (an important group as they do not smoke cigarettes) which suggests that long-term exposure to ozone might increase the risk of asthma in men but not in women. More work is needed to confirm or refute this finding and whether the absolute level of pollutants or the day to day increments are important generators of chronic inflammatory changes.

3.2.2 Challenge studies. The vast majority of work on bronchial challenge studies have come from the USA. Such studies have usually incorporated single pollutant exposures, either with or without exercise, usually on a static bicycle, in challenge chambers for periods of up to six hours but more usually up to two. There have been some studies of combinations of pollutant exposures, notably ozone with other gaseous pollutants. Particle challenge is technically difficult. There are some projects ongoing which are attempting to look at exposures to diesel exhausts from a static diesel rig but the bulk of published work on particle challenge considers the soluble particulate fraction, namely sulfuric acid, ammonium sulfate and ammonium bisulfate.

Sulfur dioxide is a known broncho-constrictor in patients with asthma, at concentrations which can be achieved during air pollution episodes. The more severe the asthma the more likely to react[16]. In contrast, normal subjects require much higher doses of sulfur dioxide before any change can be seen in any lung function parameter, showing very small changes at a 1000 ppb but clear evidence of effects at 4000 ppb[17]. These exposures are orders of magnitude greater than normal ambient levels and indeed, during air pollution episodes.

Nitrogen dioxide, however, has quite a different pattern of response. In challenge studies only extremely high concentrations of nitrogen dioxide have shown any change in lung function in either asthmatic or normal subjects[18]. This is somewhat in contrast to the findings from the epidemiological work where a number of studies have shown associations with variations in ambient nitrogen dioxide at levels of the order of 30 - 100 ppb[19]. This raises the question as to whether nitrogen dioxide may be acting indirectly rather than as a direct effect on the airways, despite the fact that it is a potent oxidant gas. Work at Birmingham Heartlands Hospital has shown that exposures to 400 ppb of nitrogen dioxide per hour can potentiate the effects of subsequent allergen challenge with the house dust mite allergen, DerP1[20]. These sort of levels of gaseous pollutant exposure are seen only during rare air pollution events but much more frequently in kitchens with gas stoves, which may help explain a recent study from Norfolk that women (but not men) who have gas stoves are more likely to have asthma than those with electric stoves[21].

Ozone is a powerful oxidant gas which is capable of causing histological changes in the bronchial mucosa in normal and asthmatic subjects at levels as low as 80 ppb, although observable health effects are not normally seen below 120 ppb[22]. Interestingly, both asthmatics and non-asthmatics seem to be equally sensitive to this gas, in marked distinction to sulfur dioxide. Again, ozone has been shown to have a potentiating effect on challenge with grass pollen[23]. Bearing in mind that both ozone and grass pollen reach their peak during the summer months, this would prove to be quite a potentially important mixture in patients who are grass pollen sensitive.

3.2.3 In vitro studies. In vitro work from the UK of cultured bronchial mucosal cell monolayers have shown that nitrogen dioxide at around 400 ppb will reduce the intactness of the bronchial mucosa making it more leaky to the ingress or egress of molecules even of significant size[24]. It is conceivable, therefore, that this would allow greater penetration of allergens or particulate pollution through the protective mucosa, should an individual be co-exposed. The same group also showed an effect on ciliary beat frequency in explanted nasal mucosal biopsies but only at much higher concentrations of nitrogen dioxide[25] (the cilia are responsible for clearing mucus from the airways). Whether this is likely to be an important factor on a day to day basis is unclear, although it is important to realise that cumulative exposure to gaseous pollutants very rapidly will exceed the total exposure in a single challenge dosing experiment.

3.3 Chronic Obstructive Airways Disease (COAD)

Patients with this condition (which incorporates chronic obstructive bronchitis, chronic productive bronchitis and emphysema, amongst other various diagnoses) are usually over the age of 50 and the vast majority (in excess of 95%) are either current smokers or have been smokers of cigarettes in the past. Patients with this condition have airways that are persistently and irreversibly narrowed, unlike patients with asthma where the airflow limitation is reversible. Possessing such narrowed airways, there is much less room for manoeuvre in maintaining their airways open so that only very small further reductions in airway diameter can cause marked falls in airflow and, therefore, increasing symptoms. Bearing in mind how common this condition is there has been surprisingly little work done in recent years on the effects of air pollution in this group of patients. There is certainly no animal model for COAD. No computer modelling work has been undertaken, nor has there been any in vitro work on tissue samples taken from patients with this condition. Challenge studies are also noticeable by their absence, which may be understandable up to a point in that these subjects start with marked airflow limitation. However, there is no real reason why more mildly affected subjects should not be studied. There still remains the problem, of course, that these patients do have quite severe lung disease and there is an ethical aspect to challenging those with more severe disease.

3.3.1 Epidemiological studies.   A panel study from Manchester in the UK[7], of patients with either asthma or COAD (and it is slightly difficult to separate out one from the other in the paper), showed that patients with COAD did vary in their symptoms on a day to day basis as air pollution levels changed. Work from the early 1960's[26], in the days of sulfur dioxide and black smoke from domestic sources, also showed that changes in sulfur dioxide correlated well with changes in symptoms in patients with chronic bronchitis. There is, therefore, evidence that on a day to day basis patients with this spectrum of conditions can show changes in symptoms as air pollution varies. The use of day to day monitoring of lung function (i.e. peak flow) is much less helpful in this group of patients as they do not exhibit the degree of variability that asthmatic patients do.

More recently, in the London 1991 air pollution event[9] (see above) hospital admissions for patients with COPD increased by 14%, although attendances to general practitioners for exacerbations of this condition did not appear to rise. This again points to the likelihood that those with the severest disease may be more affected by changes in air pollution and that those with less severe degrees of airway narrowing show much less likelihood of symptoms increasing during periods of high air pollution. While consistent associations have been demonstrated between air pollution changes and hospital admissions and mortality for COPD in the series of studies from the USA[13,27], there are only limited data from the UK.

A series of studies from the United States by Schwartz (summarised in Ref 27), have been fundamental in reassessing the approach to the health effects of air pollution. Although the majority of his research has considered health effects in American cities (the only UK data which has so far attracted his attention being the data from the 1952 London fog incident) this series of papers has been so consistent in its findings and has produced such discussion that they must be considered here. The most important features of Schwartz' work are his use of Poisson regression analysis and his accounting for as many confounders as possible. The main confounders in studies of this type are meteorological variables, (particularly changes in temperature), social class structure and other demographic features and other co-pollutants, ensuring that all are entered into the analysis. He has refined his methods in response to criticism about inadequate confounding but his findings have remained essentially unchanged and it would appear that his approach is robust. His methodology still, however, remains open to the criticism that, in health terms, air pollutants act independently of each other and, perhaps, should not be regarded as independent variables. However, his assumption of independent action is probably acceptable as the only way that we can begin to determine which pollutant variables may, in fact, be acting together, is by performing a study such as this where they are regarded in the first instance as independent.

The results from his studies are striking, revealing that the most important air pollutant appears to be particles. Some of the studies have used older methods of measuring particle levels and so some assumptions have to be made in those cases, for instance, making arbitrary decisions as to the numeric equivalence of, say black smoke in terms of PM_{10}. Nevertheless, he has shown that for every 10 $\mu g/m^3$ rise in daily PM_{10} a 1% increase in all-cause mortality will follow, with a slightly higher rise when considering respiratory mortality. The same consistent pattern is seen for hospital admissions, particularly for COPD but also for ischaemic heart disease. The average rise in hospital admissions for respiratory disease is 1% for a 10 $\mu g/m^3$ rise and for ischaemic heart disease 1.4% for the same increment. When these data were first published they were quickly picked up by pressure groups and in the media in the UK. Attempts were made to extrapolate the findings to situations in this country even though there are clear qualitative differences in PM_{10} between the two countries (and indeed within the two countries) and the fact that meteorological variation is also different. However, it has to be accepted that

Schwartz' studies have covered a fairly wide range of climatic situations and that the broad principle can be extrapolated to the UK, although probably not the size of any effect.

On the basis of this, two studies have now been performed in the UK using the same methodology as Schwartz. The first study from Birmingham[28] confirmed an association between PM_{10} and hospital admissions for COPD. It would appear that there was a dose-response effect in that the greater the increments in PM_{10} from the previous day, the greater the number of hospital admissions. There was the impression of a lagged effect and the best fit seemed to be with a three day moving average, which would make biological sense. However, the effects were small and it was calculated that if one was able to peg PM_{10} so that it would not exceed 50 µg/m^3 on any day in a year this would save one hospital admission per fortnight in Birmingham, which has a population of one million. This study was seminal in helping the Expert Panel On Air Quality Standards (EPAQS) to arrive at their Standard for particulate pollution, namely a 50 µg/m^3 twenty four hour mean value. Interestingly, this study showed absolutely no correlation with ischaemic heart disease admissions but there was an association with hospital admissions for acute stroke. However, this latter was clearly a same day effect with no lagged associations. Subsequently, an analysis from London[29] has shown an association of mortality with ozone, in contrast to the Birmingham study. They estimated that there was a 3.5% increase in all cause mortality for a 29 ppb rise (the difference between the 10th and 90th centiles) in ozone, the respective increases for respiratory and cardiovascular mortality being 5.4% and 3.6% respectively. These findings are not only of interest but are of importance. They appear to confirm an association, albeit at low level, with mortality in hospital admissions but the findings are far from consistent and show differences from the US studies. Clearly more studies need to be undertaken in different cities and different countries.

4 Air Pollution and Heart Disease

In general terms heart disease can be divided into three broad groups. The commonest type is coronary artery disease in which the coronary arteries supplying blood to the myocardium are narrowed. This gives rise to the symptom of angina (chest pain on exertion). Occlusion of a coronary vessel leads to a myocardial infarction or heart attack. The two other groups are valvular heart disease and disease of the heart muscle, but these two groups are of no relevance as far as air pollution in concerned.

It is surprising that heart disease appears to be affected by changes in air pollution. It is well recognised that meteorological variations, particularly falls in temperature, can bring on anginal episodes and, indeed, frank myocardial infarction, but the American hospital admission and mortality studies have raised questions about whether air pollution, more specifically airborne particulates, may be playing a role.

4.1 Cardiac Effects of Air Pollution

Schwartz' series of studies has shown, as described above, a most remarkable consistency of a 1% increase in all-cause mortality for a 10 μg/m rise in PM_{10}[13]. This is seen also in patients admitted or dying from coronary disease, the increase being of the same general size. In the USA this finding is very consistent but the picture is much less clear when looking at European studies. Ischaemic heart disease deaths in the Birmingham study[28] showed absolutely no association with changes in PM_{10}, although studies from London[29] suggested that there was an association with a 2.5% increase in mortality for a 12.5 μg/m^3 rise in black smoke. However, if one accepts that the association between rises in particulate pollution and ischaemic heart disease morbidity and mortality is at least a relatively consistent finding, we are left with a difficult problem in terms of explaining how this might happen at a mechanistic level.

4.2 Potential Mechanisms

One can quite happily understand that inhaling particulate pollution might have an effect on patients with lung disease. The difficulty rests with how it could impact on patients with heart disease. The main criticism of studies which have identified an impact on heart disease has been that inadequate account has been taken of confounders which, if properly assessed, would remove any association. This however, seems not to be the case as, despite some fairly vigorous discussion, the general belief now is that this association is a true one. Seaton and colleague[30] have hypothesised that the mechanism could be due to penetration of ultrafine particles right down to the alveoli and then for the particles to become interstitialised, i.e. penetrate into the lung stroma surrounding the alveoli and terminal bronchioles. Once there, they set up an inflammatory reaction which may have an impact on blood flowing through the capillaries in the lungs, perhaps by activation of fibrinogen, one of the clotting factors. The blood, thus altered would leave the lungs and travel to the heart and may just have a sufficient impact on a compromised coronary circulation to produce symptoms or frank infarction. This attractive hypothesis is supported by animal work where interstitialization of 100 nm particles has been shown[31]. In another series of experiments[32], in which resuspended PM_{10} was introduced by inhalation into guinea pigs the untreated PM_{10} caused quite a marked inflammatory reaction in the intestitium of the lung, but if the PM_{10} was washed beforehand, then no such inflammatory reaction occurred. This would suggest that the carbon core of insoluble PM_{10} is inert (as one would expect), but that substances carried on the surface of the particles (for instance acid radicals, heavy metals, even fragments of allergen) might be the cause of this inflammatory reaction. This is an attractive hypothesis and fits with the particle surface area idea expressed above. This theory is also testable, although it is likely to be a little time before studies in man can be devised to do so, and animal studies will have to suffice in the meantime.

5 Public Health Effects

So we are left with two main problems. How much does air pollution initiate disease and how much does it exacerbate disease. From the UK viewpoint there are no longitudinal studies to assess whether current levels of air pollution are sufficient to initiate disease in patients who are previously healthy. The longitudinal studies from the United States have suggested that there may be a long-term effect. The Seventh Day Adventists study[15] implicates ozone and particulates in men and the Six Cities study[33] and the American Cancer Association study implicate particulate pollution. However, neither of these two latter studies were able to take account of childhood exposure to air pollutants as a primer for the development of problems in later life. There is now abundant evidence to suggest that childhood experience of disease, particularly viral infections, will predict the development of COPD in adult life[34] and it would seem intuitively logical that the same might apply to air pollution were adequate data sets available to assess this. One of the main problems we have in this area is a lack of studies but in particular, a lack of studies which take account of individual characteristics and personal exposures to other potential causes of respiratory disease such allergens and passive cigarette smoking. If there is a chronic effect, it is likely to be small, otherwise there would be an obvious difference between urban and rural areas in various diseases. When one considers the fact that the highest prevalence of childhood asthma in the UK is on the Isle of Skye[35] one has to reflect that air pollution effects are likely to be small in this particular area of concern.

The data on exacerbation of disease is much clearer, but as we have seen, the effects are small. Recovery is usually quick (within the day) and the impacts may not even be noticeable on an individual basis at the symptomatic level. Whether repeated exposures of this type lead to acceleration of pre-existing disease is another area of concern. This might, perhaps, be considered a semantic difference from initiation of disease but one that, nevertheless, is of great import. There are a number of cases of occupational asthma where continued exposure to an occupational agent (for instance tetrachlorophthalic anhydrides) can lead to progressive accelerating disease, even when the exposure is removed[36]. There appears to be a window of opportunity during which time one has to act to remove the individual from the source before a progression occurs. It is no great logic leap to extend this concept to ambient air pollutant exposure but at present there is no evidence to suggest that this is the case, attractive though it is.

Although these effects, where we can measure them, are small, this does not mean to say that the public health effect is small. For instance, compare the effects of cigarette smoking on the incidence of lung cancer and of ischaemic heart disease. In smokers, the incidence of lung cancer is 2.27/thousand per annum and in non-smokers 0.07/thousand per annum[37]. The ratio of these two rates give us an idea of the strength of such an effect, in this case 32.4, showing that cigarette smoking is a very powerful cause of lung cancer. When one considers ischaemic heart disease in the same way, the rate in smokers is 9.93 and in non-smokers 7.32. Consequently, the ratio between the rate in smokers and non-smokers is much smaller at 1.4 showing that, although cigarette smoke has a contributory role in ischaemic heart disease its effect is nothing like as strong as that for lung cancer. Neither of these values, however, give an indication of the effect on public health. To try and determine this, one considers the *difference* between rates in smokers and non-smokers which works out at 2.2 for lung cancer and 2.61 for ischaemic heart disease, in fact, very similar values. Consequently, a small impact affecting a large number of people can have as great an effect on the public health as a strong effect on smaller numbers of people and it is this principle that has to be borne in mind before dismissing out of hand (which would be most unwise) these health effects.

If one, therefore, accepts the health effects of air pollution, one is left with how to define this in economic terms. At present there have been no adequate assessments of the economic costs of air pollution. This is an extraordinarily complex area as a balance has to be made between the health costs and the costs of control measures. Bearing in mind the Government's approach to controlling cigarette consumption by refusing to ban the advertising of cigarettes and where the costs of losing all the revenue from cigarettes would vastly outweigh the savings made from a loss of cigarette-induced disease (at least according to Government figures), one perhaps will be forgiven for feeling slightly sceptical about whether an adequate response will be made. However, there has been a sea change in the Government's approach to the health effects of air pollution, including increasing available funding for research, which is much welcomed.

6 Summary

Air pollution in the UK causes health effects although these are of small degree. Day to day changes in ambient levels of pollution (particularly particles and ozone) cause short-lived falls in lung function - and increases in symptoms in susceptible groups, of the order of 3% on high pollution days. The susceptible represent those with pre-existing respiratory or coronary heart disease and those who are at the severe end of the spectrum. The same severity groups also appear to be at increased risk from hospital admission for an exacerbation of their condition or from dying earlier than expected. However, we do not know whether these deaths are brought forward by days, months or years. Normal subjects do not appear to suffer ill-effects on a day-to-day basis.

Whether long-term exposure to current ambient air pollution is more likely to cause healthy individuals to develop respiratory disease is not known, although it is the possibility of such chronic effects which is of justifiable concern because of the implications on public health as a whole. Long-term effects of exposure to ambient benzene and 1,3-butadiene, while in theory contributing to the cancer load, has an immeasurably small effect [38].

We do not know about trends in health load from pollution. At present, we are just beginning to identify what proportion of cancer morbidity and mortality is due to pollution. It is likely that the trend will be static, bearing in mind the relatively small measured effects and the past and projected trends in pollutant emissions, but only the appropriate prospective studies will enable us to quantify this important area.

References

1. W.P. Logan. 1952. *Lancet*, 1953, **264**, 336.
2. D.A.K. Black, J.D. Pole. *Brit. J. Prev. Soc. Med.*, 1975, **29**, 222.
3. J.G. Ayres, S. Pansari, P. Weller, A. Sykes, J. Williams, N. Butler, D. Low. *Respir. Med.*, 1992, **86**, 403.
4. D. Strachan. Epidemiology of COPD: a British perspective. In:Chronic Obstructive Pulmonary Disease. Ed. Calverley P & Pride N. Chapman & Hall, 1996, p. 47.
5. Committee on the Medical Effects of Air Pollution. London. *HMSO*, 1995.
6. P.L. Kinney, G.D. Thurston, M. Raizenne. *Environ. Health Perspec.*, 1996, **104**, 170.
7. B.G. Higgins, H.C. Francis, C.J. Warburton, C. Yates, A.M. Fletcher, C.A.C. Pickering, A.A. Woodcock. *Thorax*, 1993, **48**, 417.
8. S. Walters, J.G. Ayres, G. Archer, R.M. Harrison. *Am. J. Respir. Crit. Care Med.*, 1994, **149**, A661.
9. H.R. Anderson, E.S. Limb, J.M. Bland, A. Ponce de Leon, D.P. Strachan, J.S.Bower. *Thorax*, 1995, **50**, 1188.
10. S .Walters, J. Miles, G. Archer, J.G. Ayres. *Thorax*, 1993, **48**, 1063.
11. A. Ponce de Leon, H.R. Anderson, J.M. Bland, D.P. Strachan, J. Bower. *J. Epidemiol. Comm. Health*, 1996, 33(Suppl 1), 563.
12. S. Walters, R.K. Griffiths, J.G. Ayres. *Thorax*, 1994, **49**, 79.
13. J. Schwartz. *Environ. Res.*, 1994, **64**, 26.
14. A.P. Verhoeff, G. Hoek, J. Schwartz, J.H. Van Wijnen. *Epidemiology*, 1996, **7**, 225.
15. D.E. Abbey, F. Petersen, P.K. Mills, W.L. Beeson. *Arch. Environ. Health*, 1993, **48**, 33.
16. D. Sheppard, A. Saisho, J.A. Nadel, A.J. Boushey. *Am. Rev. Respir. Dis.*, 1981, **123**, 486.
17. R.W. Stacy, E. Seal, D.E. House, J. Green, L.J. Rodger, L. Raggio. *Arch. Environ. Health* 1983, **38**,104.
18. W.S. Linn, D.A. Shamoo, E.L. Avol, G.D. Whynot, K.R. Anderson, T.G. Venet, J.D. Hackney. *Arch. Environ. Health*, 1986, **41**, 292.
19. M. Rutishauser, U. Achermann-Liebrich, C.H. Braun, H.P. Gnehm, H.U. Wanner. *Lung*, 1990, **168**(Suppl), 347.
20. W.S. Tunnicliffe, P.S. Burge, J.G. Ayres. *Lancet*, 1994, **344**, 1733.
21. D. Jarvis, S. Chinn, C. Luczynska, P. Burney. *Lancet*, 1996, **347**, 426.
22. D.H. Hortsman, L.J. Folinsbee, P.J. Ives, S. Abdul-Salaam, W.F. McDonnell. *Am. Rev. Respir. Dis.*, 1990, **142**, 1158.
23. N.A. Molfino, F.C. Wright, I. Katz, S. Tarlo, F. Silverman, P.A. McClean, J.P. Szalai, M. Raizenne, A.S. Slutsky, N. Zamel. *Lancet*, 1991, **338**, 199.

24. J.L. Devalia, C. Rusznak, R.J. Davies. *Respir. Med.*, 1994, 88, 241.
25. J.L. Devalia., R.J. Sapsford, D.R. Cundell, C, Ruznack, A,M, Campbell, R, Davies. *Eur. Resp. J.*, 1993, 6, 1308.
26. P.J. Lawther, R.E. Waller, M. Henderson. *Thorax*, 1970, 25 525.
27. D.W. Cockery, C.A.P ope. *Annu. Rev. Pub. Health, 1994, 15, 107.*
28. J. Wordley, S. Walters, J.G. Ayres. *Thorax,* 1995, 50, A34.
29. H.R. Anderson, A. Ponce de Leon, J.M. Bland, J.S. Bower, D.P. Strachan. *Brit. Med. J.,* 1996, 312, 665.
30. A. Seaton, W. MacNee, K. Donaldson, D. Godden. *Lancet,* 1995, 345, 176.
31. J. Ferin, G. Oberdorster, D.P. Penney. *Am. J. Respir. Cell. Mol. Biol.*, 1992, 6, 35.
32. J.S. Tepper, J.R. Lehman, D.W. Winsett, D.L. Costa, A.O. Ghio. *Am. J. Resp. Crit. Med.*, 1994, 149, A839.
33. D.W. Dockery, C.A. Pope, X. Xu, J.D. Spengler, J.H. Ware, M.E. Fay et al. *New Engl. J. Med.*, 1993, 329, 1753.
34. D.J.P. Barker, K.M. Godfrey, C. Fall, C. Osmond, P.D. Walker, S.O. Shaheen. *Brit. Med. J.*, 1991, 303, 671.
35. J.B. Austin, G. Russell, M.G. Adam, D. Mackintosh, S. Kelsey, D.F. Peck. *Arch. Dis. Child,* 1994, 71, 211.
36. K.M. Venables, M.D. Topping, A.J. Nunn, W. Howe, A.J. Newman Taylor. *J. Allergy Clin. Immunol.*, 1987, 80, 212.
37. R. Doll, A.B. Hill. *Brit. Med. J.,* 1964, 1, 1399.
38. Expert Panel on Air Quality Standards. Benzene. *HMSO,* 1995.

Current and Future Legislation - United Kingdom and Europe

M.L. Williams

Department of the Environment, 43 Marsham Street, London SW1P 3PY, UK

1 Introduction

The quality of the air we breathe is of critical importance to everyone. Yet we all aspire to higher standards of living, which entails a growth in demand for industrial production and for mobility. It is therefore necessary to find a way in which meeting those demands can be combined with achieving and maintaining a clean and healthy environment in which individuals and communities can thrive. This was recognised in both Agenda 21, the touchstone document to come out of the Rio Earth Summit in 1992, and the UK's own Sustainable Development Strategy.

The air in the UK is significantly cleaner today than in the 1950s and 1960s and levels of air pollutants are expected to fall dramatically over the next decade. However, while healthy individuals are now unlikely to experience acute effects at typical air pollution levels, there is some evidence of associations with advanced mortality, chronic illness and discomfort for sensitive groups. In some local areas - particularly congested urban centres - emissions from traffic, industry and other sources can still affect the quality of life for all.

A strategy is therefore needed to improve areas of poor air quality, to reduce any remaining significant risks to health, and to achieve the wider objectives of sustainable development in relation to air quality in the United Kingdom. The Environment Act 1995 included a requirement for the development of such a strategy; provided for the further development of local air quality assessment and management; and provided new regulatory powers for the improvement of air quality. This paper summarises the National Air Quality Strategy which has been issued for consultation. A finalised version of the strategy, duly amended, was adopted at the end of 1996.

2 Overview

The Government intends, through this Strategy, to provide a clear and work-
able framework for improving air quality. It has therefore been based on the
following principles:

* A statement of the Government's general aims for improving air quality;
* Clear measurable targets. These are based, as far as possible, on an
  understanding of the health effects of the pollutants concerned and the
  costs of feasible abatement methods.They have identified timescales for
  their achievement;
* A balance between national and local action, to ensure a flexible and
  cost- effective approach to air quality management;
* A transparent framework to allow all parties - business, local government
  and the wider community - to identify the contribution they can make to
  better air quality;
* The need to ensure consistency with international commitments; and
* Regular review of the Strategy and all the elements contained therein.

The *key elements* of the Strategy are:

* Health-based air quality standards and objectives, to act as reference
  points by which policies are to be directed;
* A target of 2005 for achievement of the objectives;
* Policies for meeting those objectives, including an assessment of the
  improvements already expected under current policies, and where and
  how those policies might need to be supplemented;
* The contribution key sectors - in particular industry, transport and local
  government - can make towards the cost-effective achievement of those
  objectives;
* A commitment to review the Strategy every three years.

At the core of the Strategy, therefore, are *air quality standards and objectives.*

Air quality standards are concentrations of pollutants in the atmosphere which
can broadly be taken to achieve a certain level of environmental quality. The
standards which are proposed are based on an assessment of the effects of each
pollutant on public health. They are taken from the best available consensual
view of medical experts. The standards will be used as benchmarks or reference
points for setting objectives. In setting standards the Government has accepted
the judgements of the Expert Panel on Air Quality Standards about the levels
of air pollutants at which there would be an extremely small or no risk to
human health. Where no recommendation has yet been made by the Panel, the
air quality standard has been derived from work by the World Health
Organization.

Air quality objectives provide the framework for determining the extent to
which policies should aim to improve air quality. They represent the Govern-

ment's best judgement of the progress which can be made in a cost-effective manner towards the air quality standards, by 2005. The **general objectives** give a broad indication of what the Government hopes will be achieved by the implementation of the Strategy and identifies the priority areas. These are complemented by **specific objectives** which are measurable and against which progress can be assessed.

It is these general and specific objectives which drive air quality policy and the implementation of Part IV of the Environment Act 1995. (This Act does not extend to Northern Ireland where parallel legislation will be introduced.) With the exception of ozone, the Government will incorporate its objectives into regulations following consultation. This means that the specific objectives contained in Table 1 will then trigger the duties of local authorities under the relevant provisions of the Act. These provisions require the development of action plans when objectives are breached or at risk. (As ozone is not generally susceptible to local abatement strategies, it will not be included in the regulations.) The standards and objectives will be subject to review in 1999, particularly those specific objectives which, because of the uncertainties attached to them, are to be treated as provisional (see below).

3 Standards and Objectives

The Government's primary objective is to ensure that all citizens behaving responsibly should have access to public places with minimum risk to their health and quality of life. The implication of this objective is that air quality policy should be directed towards getting and keeping air quality as close to the benchmark standards proposed here as is reasonable and justifiable on consideration of the costs and benefits. Table 1 below summarises the air quality standards and specific objectives which the Government proposes to incorporate in its air quality strategy. The averaging periods reflect the differing effects on human health: standards for pollutants which have acute effects have short averaging periods.

For those pollutants which are starred in the table below, the objectives will be met in most places. However, present estimates point to a gap between the reductions required to meet the objective levels everywhere and those likely to be achieved by measures adopted or announced by the Government or likely to be introduced on a European basis. These objectives are therefore to be treated as provisional, being more likely than the other objectives to be changed when the strategy is reviewed in 1999 - or, if compelling evidence becomes available sooner, prior to that review. Unless and until they are changed, however, they are to be treated in just the same way as the other objectives.

The 1999 review will need to address the gap remaining between the improvements necessary and those provided. The size of that gap will depend on factors which are difficult to quantify at this stage, for example: the weather; the accuracy of emissions forecasts; and the extent to which emissions

Table 1   *Proposed standards and specific objectives*

Pollutant	Standard		Specific objective to be achieved by 2005
	concentration	measured as	
Benzene	5 ppb	running annual mean	the air quality standard
1,3-Butadiene	1 ppb	running annual mean	the air quality standard
Carbon monoxide	10 ppm	running 8-hour mean	the air quality standard
Lead	0.5 $\mu g/m^3$	annual mean	the air quality standard
Nitrogen dioxide	104.6 ppb	1 hour mean	104.6 ppb, measured as the 99.9th percentile, to be
	20 ppb	annual mean	achieved by 2005*
Ozone	50 ppb	running 8-hour mean	50 ppb, measured as the 97th percentile, to be achieved by 2005*
Particles - PM_{10}	50 $\mu g/m^3$	running 24-hour mean	50 $\mu g/m^3$ measured as the 99th percentile, to be achieved by 2005*
Sulfur dioxide	100 ppb	15-minute mean	100 ppb measured as the 99.9th percentile, to be achieved by 2005*

ppm = parts per million; ppb = parts per billion; $\mu g/m^3$ = micrograms per cubic metre
* these objectives are to be treated as provisional, as described above.

reductions are available through reasonable local action (such as tough enforcement and traffic management measures). The review will also need to take into account other relevant factors, such as consumer behaviour (particularly driving patterns), advances in scientific or medical understanding of the effects of air pollution, and further analysis of the costs and benefits associated with the measures needed to achieve the objectives.

Percentiles. In some cases, it would not be appropriate to ensure 100% compliance with the air quality objective level. This may arise for a number of reasons, eg: uncontrollable adverse weather conditions; national festivals such as Bonfire Night, which it would be inappropriate to ban or control; or disproportionate cost of feasible abatement methods. Where a specific objective does not require absolute compliance with the air quality standard, and the averaging time is short, the specific objective is expressed in terms of percentile compliance, which allows for a small number of readings in a given year to be discounted.

Specific objectives should apply in areas where people are likely to be exposed over the relevant averaging period. Policies which aim at achieving air quality objectives at points where the highest measurable concentrations prevail with no regard for whether or not people might be exposed would be inappropriate and highly inefficient.

4 The Contribution to Air Pollution from Different Sectors

Table 2 shows the relative contribution of different sectors to total national emissions of the pollutants covered in this strategy (ozone is not emitted directly - the pollutants which lead to its formation, nitrogen oxides and volatile organic compounds, are given). It must be remembered that these figures do not necessarily reflect the relative contribution of these sectors to any particular area, including pollution 'hot spots' or problem areas. There is a great deal of variation between urban and rural areas, and between residential, commercial and industrial areas.

The main sources of airborne pollution are:

Road transport. Road transport is a significant and in most urban areas the main source of emissions of all of the pollutants covered by the Strategy with the exception of sulfur dioxide. The proportion of emission varies depending on the traffic, weather and other local sources.

Energy generation. Combustion plants which provide public power are currently the dominant source of sulfur dioxide, and also produce nitrogen oxides, some particles and other pollutants. Smaller combustion plants have less impact at a national level but can affect local air quality, again contributing towards emissions of sulfur dioxide, nitrogen oxides and particles.

Industrial processes. Many different sources are involved. For example:

◆ Volatile organic compounds evaporate from many liquid fuels, paints and cleaners and are also formed during combustion processes.
◆ Lead is widely used to make batteries, and is also used in pigments, lead piping, radiation shielding and tanks.
◆ Particles are given off by quarrying, construction and many other industrial processes.

Table 2   *The main sources of pollutants - national emission figures*

Pollutant source	PM_{10} 1994	NOx 1993	CO 1993	SO_2 1993	Lead 1994	Benzene 1993	VOC 1994	Butadiene 1993
Transport	29%	59%	90%	5%	64%	68%	35%	77%
Combustion (Industry)	35%	24%	3%	92%	4%	5%	2%	–
Chemicals and Fuels	–	–	1%	–	–	11%	25%	18%
Other Industrial	24%	14%	–	–	27%	16%	25%	–
Waste	–	–	1%	–	4%	–	–	5%
Domestic/ Other	14%	3%	5%	3%	1%	–	13%	–

Domestic sources. Emissions from domestic fires have reduced significantly at a national level as a source of pollution, as local smoke control areas have been introduced. However, some areas remain where domestic emissions of sulfur dioxide and particles have a significant impact on air quality. In addition, domestic use of solvents in paints, varnishes and other products is an important source of volatile organic compounds.

5 Progress from Current Policies

Emissions of airborne pollutants in the United Kingdom are subject to a wide range of controls, including:

Smoke Control Areas - Clean Air legislation allows local authorities to initiate smoke control programmes. With a few exceptions, only smokeless fuels may be burnt in these areas. This control, alongside the wider availability of gas and electricity, led to the large scale reduction in emissions of sulfur dioxide from domestic coal-fires.

Authorisations for specified industrial processes - 2,000 processes with the potential for polluting a number of environmental media are regulated under the system of Integrated Pollution Control (IPC). (A similar system of control will be introduced into Northern Ireland under the Industrial Pollution Control (Northern Ireland) Order currently in preparation.) Operators are required to adopt the Best Practicable Environmental Option (BPEO) for controlling emissions to air, water and land. Some 13,000 further processes are regulated under Local Air Pollution Control (LAPC). IPC and LAPC are both based on the principle of Best Available Techniques Not Entailing Excessive Cost (BATNEEC). Authorisations are agreed on a plant-by-plant basis allowing account to be taken of local circumstances. The first reviews of IPC authorisations have recently been completed in England and Wales, covering power stations. Considerable reductions in emissions have been required. Further large reductions are expected from those power stations which are changing to combined cycle gas turbine stations.

Emissions standards for fuels and new road vehicles - Although European vehicle emission and fuel quality standards were first introduced in the early 1970s, until the late 80s the increase in road transport was accompanied by similar increases in pollution. Standards agreed in the European Community since the early 90s have reduced emissions from new vehicles substantially. For example, the 1993 standards have meant that new petrol-engined cars must be fitted with a catalytic converter. Converters, if fully functioning, can reduce emissions by up to 80%. Over the next ten years the fleet will be largely replaced by vehicles conforming to these much tougher standards, which should lead to dramatic falls in vehicle emissions. If road traffic continues to increase, however, these gains could be jeopardised. The other substantial improvement has come from the reduced use of lead additives in petrol. Recent surveys indicate that blood lead levels have decreased by around 50%

since the mid-eighties. A significant contributory factor to this is the reduction of the maximum lead allowed in petrol, coupled with the increased use of unleaded fuel, encouraged by fiscal incentives. These measures have reduced vehicle emissions of lead by around 80% since 1975.

6 What Current Policies are Likely to Achieve

For some of the traffic related pollutants, notably **benzene, 1,3-butadiene, carbon monoxide** and **lead**, the continuing improvements from vehicle and fuel standards already in place are expected to be sufficient to bring ambient concentrations across the UK down to the proposed objective levels by 2005.

Nitrogen dioxide and particles are both essentially urban pollution problems. In order to achieve the objective for those pollutants, it is estimated that urban emissions need to fall by around 60% from 1995 levels. Policies already in place will not be sufficient on their own to achieve this scale of emission reduction.

A similar story is true for **sulfur dioxide**. The new settlement for English and Welsh power generation plants agreed with HMIP (now part of the Environment Agency), in securing dramatic reductions in sulfur emissions over the next decade, goes a long way to achieving the objective levels. It is likely that the risk of exceedences will remain in those few areas where domestic coal burning is still prevalent, or where there is a concentration of small/medium sized combustion sources.

For **ozone**, the Government has set an ambitious target, which, due to the transboundary nature of the problem, can not realistically be achieved without the cooperation of neighbouring countries in northwest Europe. Estimates vary on the size of reduction needed, *across the region*, in emissions of ozone's precursor pollutants; nitrogen oxides and volatile organic compounds. These estimates range from around 50-70% on current levels, which will not be achieved on current policies alone. The Government will continue to press for coordinated action on ozone in Europe, as agreed at the London Ministerial Conference on ground-level ozone in April 1996.

7 Meeting the Objectives

Considerable measures are already being taken to improve air quality, and the reductions which are expected as a result are dramatic. Nevertheless, substantial further progress will be required to meet the objectives by 2005. More stringent standards for new vehicle and fuel standards will be particularly important; as will the continued application of BATNEEC to industrial processes. Other national measures may also be necessary, and targeted local action will be needed to ensure a flexible and cost-effective approach is taken towards tackling local hotspots.

Transport. The European Commission has recently brought forward proposals for further, more stringent vehicle and fuel standards for the year 2000 and 2005. Current indications are that the proposed standards for 2000 will secure 50-55% reduction on 1995 emissions. This means that there may be a shortfall, leaving some areas short of meeting the objectives. More stringent standards which may be adopted for 2005 will not have effect within the time frame of this strategy, but the final settlement of any standards for the year 2005 will have a significant bearing on the measures required to maintain the level of air quality improvement achieved by that date, against the likely growth in traffic.

In some areas further measures may be needed to make up the possible shortfall predicted for 2005 and so to meet the specific objectives. The policy gaps associated with both PM_{10} and NO_2 are currently estimated at around 5-10% in 2005 and it is not yet clear whether the measures available to reduce levels of these pollutants will be capable of meeting the objectives cost-effectively. The objectives for these pollutants and the uncertainties of meeting them are discussed above.) Potential measures include:

- tighter controls on the existing vehicle fleet, its management and operation;
- development of environmental responsibilities by fleet operators, particularly public service fleet operators, and by the public at large, in transport and vehicle use; and
- changes in planning and transport policies which would reduce the need to travel and reliance on the car.

The Government will be issuing draft guidance on traffic management later this year. Other measures include regulations giving local authorities powers to conduct, with the police, roadside emissions testing; and regulations allowing local authorities to prevent drivers in parked vehicles from leaving their engines running unnecessarily.

The Government also believes that fiscal incentives can have a role to play in achieving environmental objectives. Previous budgets have included measures on fuel duty in this regard, and the Government will keep under review the potential for fiscal incentives to encourage further reductions in vehicle emissions.

Industry. The other major reduction in emissions is expected from the four-yearly reviews of IPC and LAPC authorisations. These reviews consider the costs and benefits of newly available technologies and good practice, and often lead to considerable reductions in allowable emissions. Most initial improvements will take place during 1996-1998. In addition, under the Environment Act 1995, the Environment Agency and the Scottish Environmental Protection Agency are required to pay regard to the strategy and the air quality standards and objectives when carrying out their functions, including granting authorisations. Fiscal measures to control industrial and other emissions will also be explored and the government will work with industry and local government on voluntary measures, and to help develop industry's role in local air quality management.

8 Local Air Quality Management

National measures must go a long way to meeting the air quality objectives. Yet, emissions vary dramatically in different areas, depending on the geography, industry and traffic. Local ownership of air quality management will be vital to ensure that solutions are cost-effective, thorough and integrated with other local needs and priorities. Local authorities are already key players. They enforce some air quality legislation. Their decisions on a wide range of issues such as planning, transport, waste and energy also have important implications for air quality. Many have asked for a wider, more strategic, role.

The Strategy encourages all local authorities to develop a more integrated approach towards local air quality management which would involve:

- ensuring an appropriate local framework, so that air quality is considered alongside issues such as transport or land-use planning;
- working with other local authorities, the highways authorities and representatives of the local community;
- building on business initiatives and establishing voluntary agreements; and
- informing and involving the public.

In the early years of the Strategy all district and unitary authorities will be required to assess and review their air quality. Where national air quality objectives are not met or are at risk, local authorities will have to designate Air Quality Management Areas and make action plans for improvements in air quality. The Government will be providing a wide range of assistance to local authorities in developing this new approach. This will include technical help on air quality assessments, new powers to test vehicle emissions, guidance on traffic management techniques for improving air quality, land-use planning and public information strategies.

9 Assessing Air Quality: Monitoring, Modelling and Inventories

Air quality assessments are a vital part of the overall Strategy, which allow us to judge whether standards will be exceeded and by how much. They also allow us to provide the public with detailed and accurate information and forecasts, via CEEFAX, TELETEXT, the Internet (http:/www.open.gov.uk/ doe/doehome.htm), a freephone telephone number (0800 556677) and through many newspapers. On a longer time-scale assessments allow us to monitor and predict the effect of current policies and help develop further national and local policies.

Monitoring. The Government is currently expanding both the urban and rural monitoring networks. To ensure the maximum cost-effectiveness, this will be done by establishing a further 10 multi-pollutant automatic urban sites during

1996 and by integrating around 35 quality assured local authority sites into the national network. New air quality monitors will be sited in areas where people are likely to be exposed to high concentrations of pollutants over the appropriate time-scales.

Emissions Inventories and Air Quality Models. Although air quality monitors are often essential they are not always the most cost-effective solution. In some situations, for example as a decision-making tool, numerical air quality modelling coupled with information about sources and background levels may be more appropriate. The Government is preparing guidance to help local authorities use the most appropriate methods of reviewing and assessing air quality.

10 Conclusion

The standards and objectives, which the Government is proposing, are ambitious; their achievement should signify the consistent attainment of very clean air throughout the country, and should form the basis for the sustainable protection of human health. The setting of the objectives is, however, a measured and proportionate response to the current effects of air pollution, and has attempted to ensure that air quality policies are directed by an appropriate consideration of the costs and benefits.

Developments in Air Pollution Measurement Techniques

C.J. Dore and J. McGinlay

AEA Technology plc, National Environmental Technology Centre, Culham, Abingdon, Oxfordshire OX14 3DB, UK

Introduction

There have been many advances in air pollution measurement techniques over recent years. The introduction of new legislation has meant that the demand for good quality monitoring data has greatly increased, and consequently several techniques have become well established through extensive use.

Several of the more complex techniques have moved from the research and development stages to becoming commercially available, and the instrumentation is now considerably more reliable. However, the greatest changes by far have been observed in the measurement techniques suited to long term monitoring. Passive and automatic point monitors are now extensively used in routine long term monitoring, rather than just for research applications. Many of the automatic point monitors now incorporate additional features, such as automatic internal calibrations, as well as giving increased reliability. Passive samplers are now available for the majority of pollutants of interest, but although the cost is usually lower than for the corresponding automatic point monitors, the lack of time resolved results makes the automatic monitors more desirable for a single site requiring high precision results. The different monitoring options are given detailed consideration in McGinlay *et al* (1996) and Bower (1997). A comprehensive list of compounds, which are of interest, accompanied by relevant information is given in the Air Quality A to Z (Bertorelli, 1995).

1 Detector Types

1.1 Passive Detectors

Passive samplers such as diffusion tubes are the cheapest method of monitoring air quality and can give a good overall picture of average pollutant levels in an area. The low cost per tube permits sampling at a number of points in the area of interest. This is useful in highlighting points of high concentration, or 'hotspots', such as major roads or emission sources, where more detailed studies may be needed.

Passive sampler surveys are becoming increasingly popular in many parts of the UK and world-wide (see Stevenson *et al*, 1997). Although results from single point passive samplers are not as precise as those from automatic point monitors, the accuracy and reproducibility of the measurements has increased over recent years. The accuracy of the results from these techniques greatly depends on the laboratory carrying out the tube preparation and analysis. Similarly, the user must take care to avoid contamination. Multiple tubes are often exposed at a site to decrease the measurement uncertainty, and avoid data loss through contamination.

Diffusion tubes are typically clear plastic tubes open, or with a membrane screen, at one end and a pollutant-absorbing chemical matrix or gel at the closed end. The tubes are prepared and sealed before being transported to the monitoring site. At the site, the tube is exposed, for example, by removing a cap and leaving it for a period of between one week and one month. The diffusion tube 'collects' the pollutant during the exposure period, at the end of which the tube is re-sealed and returned to an analytical laboratory. The quantity of pollutant absorbed enables an estimate of the average ambient pollutant concentration over the exposure period to be obtained.

The time resolution of this technique is limited, as it can only provide information on integrated average pollutant concentrations over the exposure period (typically 1-4 weeks). Many air quality standards and guidelines are based on short term measurements (hourly or daily averages) and compliance with these can only, therefore, be determined directly using automatic monitors. For some applications, however, statistical techniques allow the likelihood of non-compliance with short-term standards to be estimated from long-term passive sampler measurements. Such estimation techniques need to be used with caution.

Passive samplers have been widely used for many years in personal exposure monitoring and occupational health assessments. For monitoring ambient air, passive samplers are particularly useful for baseline surveys, area screening or indicative monitoring. They can also be useful when used in combination with active samplers or automatic analysers. In such hybrid surveys, passive samplers can provide geographically resolved air quality data, whilst the more sophisticated devices offer time-resolved information on concentration peaks and diurnal variations. Hybrid surveys of this type can be particularly cost-effective.

Passive samplers require no power or other services and can simply be mounted on the sides of buildings, or onto covered shelters or street furniture such as lamp-posts and railings. The tubes must be held vertically with the open end downwards during sampling. Generally a permanent clip is mounted so that the tubes can be changed easily. The clip and spacer (e.g. block of wood or plastic) may be simply mounted at the monitoring site with PVC tape, double sided tape, or cable tie as appropriate.

It is important that the tube is situated in a generally open area, which allows free circulation of air at the open end of the tube. Also, samplers must not be mounted directly onto a surface since absorption of the pollutant species by the surface may lead to a thin layer of reduced atmospheric concentrations immediately adjacent. To avoid this problem the tubes are mounted using a spacer block of wood or plastic, with the open end of the tube located below the lower surface of the spacer.

The samplers must be uniquely identified and careful records maintained of the location of each sampler and the duration of exposure. In addition, full details of each site including location, grid reference and full description must be kept for the subsequent analysis. In some cases it is recommended that replicate tubes be used when a single measurement is important. Replication of measurements is also useful for quality assurance and quality control (QA/QC) intercomparisons (see Chapter 8).

In some cases, measurements of one pollutant may show a close correlation with other pollutants. Preliminary work suggests that, for screening studies of traffic pollution, an estimate of CO and PM_{10} levels can be determined from NO_2 levels obtained from diffusion tube studies. Similarly, an estimate of 1,3-butadiene levels may be obtained from ambient benzene levels. However, this gives estimates only, and further measurement is usually recommended.

1.2 Active Samplers

Active sampler methods collect pollutant samples either by physical or chemical means for subsequent analysis in a laboratory. Typically, a known volume of air is pumped through a collector such as a filter or chemical solution for a known period of time, which is then removed for analysis. Samples can be taken each day, thereby providing reasonable time resolution, but at a relatively modest capital cost compared with automatic monitoring methods. The principle active sampler methods used in the UK are those employed to measure black smoke and SO_2, but active samplers are also used to measure lead and other trace elements in air. There is a long history of sample measurements in the UK and throughout Europe, providing valuable baseline data for trend analyses and comparison. For this reason, these measurement techniques have found continued and extensive use, even though automatic monitors are becoming increasingly popular.

1.3 Automatic Real Time Point Monitors

These produce high resolution measurements (typically hourly averages or better) at a single point for pollutants such as ozone, oxides of nitrogen, sulfur dioxide, carbon monoxide and PM_{10} particulates. Gas Chromatography (GC) analysers also provide high resolution data on benzene, 1,3-butadiene and other speciated hydrocarbon concentrations. The sample is analysed on-line and in real-time. Although this is the most expensive method of air quality monitoring routinely employed, the number of instruments being used is rapidly increasing. Automatic monitors are used in the UK automatic urban, rural and hydrocarbon networks (Bower *et al*, 1996). In order to ensure that the data produced are accurate and reliable, high standards of maintenance, operational and QA/QC procedures are often required. This is a direct consequence of producing on-line data, rather than the laboratory conducting the analysis of passive or active samples and being responsible for a large part of the QA/QC.

Real time point monitors are usually termed 'automatic' or 'continuous'. Automatic methods utilise proven high performance technology to provide time resolved data. However, the high resolution of data is accompanied by increased cost and complexity, with lower operational reliability. Several of the instruments, most notably NO_x, SO_2 and O_3 detectors, are now suitably established and their reliability has greatly increased (although routine maintenance and calibration is still required).

In addition to the monitor itself, a variety of other ancillary equipment is also required for instruments. This might include data handling facilities, calibration gases and site temperature regulation, as well as a secure location with all the necessary utilities (e.g. electricity, telephone) for the equipment. Trained and skilled manpower is also required to operate such stations, with regular site visits and maintenance being essential. Furthermore, continuous analysers also produce large quantities of data, which usually necessitate telemetry systems for data acquisition and computers for subsequent processing and analysis.

Despite the sophistication of the equipment, the quality of the data produced is only as good as the QA/QC system employed at each measurement site. Automatic analysers require regular calibration with traceable gas standards. Effective screening of data to detect equipment faults or other problems is also required to assure the quality of the resulting data and ensure the measurements are credible and reliable. Chapter 8 considers the appropriate QA/QC programmes and some data handling issues for automatic analysers. Calibration of several different types of these point monitors uses an internal calibration source, and zero air. It is usually possible to set the instrument to automatically conduct a zero and span calibration at regular intervals.

1.4 Long Path and Spatially Resolved Monitoring

In recent years, long-path optical remote sensors have become commercially available for ambient monitoring. Due to the expense and complexity these detectors are usually used for short-term measurement campaigns rather than long-term monitoring programmes. These long-path instruments measure the average concentrations of several pollutants along an open path, typically of the order of a few hundred metres. Ultraviolet (UV), visible and infrared (IR) radiation from a lamp is directed at a receiver or retroreflector several hundred metres away. The spectrum of the light received from the lamp contains information about the concentrations of the pollutant gases through which the light has passed (Grant *et al*, 1992). The data obtained therefore consist of the average pollutant gas concentration along the light path. Some techniques enable range-resolved information to be obtained.

These methods usually offers the advantage of being able to measure several different pollutants conveniently in one system, with no direct contact with the sample gas. However, the technology is often expensive. Careful consideration also needs to be given to the calibration and quality control of long-path monitors. Some methods are still under development, and it is expected that improvements to these systems will continue for some time, until operation becomes less specialised. A disadvantage of some of these techniques, as with all optical remote sensing methods, is that the measurement may be lost or degraded during low visibility weather conditions such as fog.

These methods are not currently appropriate for monitoring compliance with EC Directive limit values and, at this time, are not currently deployed in the UK National Monitoring Networks. However, they can be powerful techniques in near-source or industrial situations, as well as in circumstances where large areas need to be scanned from a single point.

1.5 Portable and Mobile Monitoring

Automatic monitoring instruments may be installed in a mobile vehicle or other portable monitoring facility, and can be used to sample whilst in motion, or deployed for monitoring campaigns lasting a few days or weeks. The main application for portable/mobile monitoring is for screening studies and to locate trouble spots ('hot-spots') where there is a need to provide monitoring data quickly. These types of study are particularly effective when carried out in conjunction with permanent fixed-point automatic monitoring studies.

A range of relatively low-cost continuous analysers have been developed for portable monitoring applications. Such analysers use electrochemical or solid-state sensor technology. These offer the advantages of real-time ambient monitoring, using small, light-weight instrumentation, but are less sensitive than conventional point monitors, and often subject to interference from other pollutant species, temperature and humidity. In some analyser types, stability of response may also be a problem and frequent calibration may be necessary.

Although potentially useful in screening studies and in identifying hot-spots, there are a number of major disadvantages to short-term monitoring studies. The principal drawback is that the resulting datasets cover a limited period which may not be temporally representative of ambient conditions, resulting in compromised data credibility and reliability. Moreover, short-term monitoring makes it impossible to observe important seasonal variations, which are significant for many pollutants. Finally, short-term studies cannot provide information on long-term trends at a particular location. Table 1 shows a summary of the advantages and disadvantages of the different techniques discussed above.

Table 1   *Advantages and disadvantages of different instrumented air monitoring techniques (McGinlay, 1996)*

Method	Advantages	Disadvantages
Passive Samplers	Very low cost Very simple Useful for baseline and screening studies	In general only provide monthly or weekly averages Laboratory analysis required
Active Samplers	Low cost Easy to operate Reliable operation/ performance Historical data set	Provide daily averages Labour intensive Laboratory analysis required
Automatic Analysers	Proven High performance Hourly data On-line information and low direct costs	Complex Expensive High skill required High recurrent costs
Remote and Long-Path Sensors	Provide path or range-resolved data Useful near sources and for vertical measurements in the atmosphere Multi-component measurements	Complex and expensive Difficult to calibrate and validate Not always comparable with conventional analysers

2 Ozone (O₃)

2.1 Passive Diffusion Tube Method for Ozone

The following section gives an overview of the preparation and analysis of ozone diffusion tubes. Compared to NO_2 and SO_2 diffusion tubes, O_3 diffusion tubes are not widely used. The measurement technique, although commercially available, has been available for a relatively short period of time. Consequently the method is not as well established, and fewer comparisons with other measurement techniques have been conducted.

Each diffusion tube consists of:

- A grid of woven wire cloth cut to a diameter sufficient to cover the end of the tube yet fit easily into the cap;
- A membrane which acts as a filter;
- Polyethylene end caps to retain the grid and the membrane;
- The tube itself, which is of FEP or PTFE with moulded or turned down ends to accept plastic caps.

All of these components are thoroughly washed in deionised water, and are then dried in an oven.

2.1.1 Preparation and Analysis.   Clean working conditions are essential with all diffusion tube preparation and analysis. Gloves, washed in deionised water, are worn throughout the preparation. Additionally all implements, such as tweezers are thoroughly cleaned in deionised water. Solutions for the adsorption solution and analysis are made with high grade deionised water, accredited pipettes and clean glassware.

The absorbing solution contains sodium nitrite, potassium carbonate and glycerol in deionised water. A known amount of the solution is placed onto a grid which is located inside an end cap. A second grid is then place over the first grid, and the tube is inserted into the end cap to secure the grids in place. A membrane is placed over the other end of the tube, and held in place by an open ended cap. As the cap is open ended the tubes are kept in sealed vials until the exposure time commences.

After exposure, the tubes are sealed in vials until extraction and analysis. Extraction is performed by adding a known volume of deionised water to the grids, and allowing the grids to stand for at least one hour to complete the extraction process.

The nitrite, from the absorption solution, and ambient ozone react to give nitrate which is quantified by ion chromatography (IC). This is corrected for reagent and travelling blanks as appropriate. This then allows the amount of ozone absorbed onto the grids to be calculated.

2.2 Automatic Point Monitor Method for Ozone

Automatic point monitors measure O_3 concentrations by UV absorption. This technique is now well established, and instruments differ little in their performance. The air sample is passed into an absorption cell, where the attenuation of UV light at wavelength 254 nm is measured by a vacuum photodiode. Each air measurement is accompanied by a reference measurement, where the air sample is first passed through an ozone-removing scrubber. This enables corrections to be made for any instability in the UV source. Typical specifications for these instruments are as follow (Bower *et al*, 1996):

Lower Detectable Limit		2 ppb
Precision as % of Upper Range Limit		<1%
Noise	Zero	1 ppb
	Span	3 ppb
24 Hour Drift	Zero	2 ppb
	Span	2% Full Scale
Linearity Error		2%
95% Response Time (max.)		150 seconds

Calibration of the automatic point monitor may be conducted by several different methods. However the majority of real time point monitors have internal calibration facilities. Zero air is produced by passing ambient air through a charcoal scrubber before entering the reaction cell. The span gas is produced by the action of UV light on the same zero air stream to produce O_3.

Further calibration techniques are possible, and these are discussed in the section on the Automatic Urban Network (AUN) (see Section 16.1).

3 Nitrogen Oxides (NO_x, NO_2, NO)

The primary constituents of NO_x are NO and NO_2. The following outlines techniques for the measurement of NO_x, NO and NO_2, either directly or indirectly. The method employed for the automatic point monitors assumes that the difference between the NO measurement and NO_x measurement is equal to the NO_2 concentrations. There are active methods for measuring NO_x, but these are not commonly used, and are consequently not included here.

3.1 Passive Diffusion Tube Method for NO_2

Diffusion tube samplers for NO_2 have, and are being, extensively used for area screening and monitoring site selection, amongst other applications, over both country-wide and local scales in various European countries. They are currently used in DoE's National NO_2 Survey at over 1200 sites throughout the UK (see Section 16.4, and Stevenson K.J., Bush T.J. and Mooney D.E., 1997). Diffusion tube surveys are also undertaken by many local authorities and private industries to assess the current state of air quality and identify 'hotspots' which may require more detailed studies. The measurement depends upon the molecular diffusion of the gas through the tube and its collection on the absorbent. 0.2 to 6 micrograms of nitrogen dioxide per sampler may be determined by this method. The NO_2 is absorbed as nitrite, and then is reacted with sulfanilamide to form a diazonium compound that couples with naphthyl ethylene diamine dihydrochloride (NEDA). This forms an azo dye. A spectro-photometric technique is then used to give quantitative values. Of all the diffusion tube methods that for NO_2 is the most established.

3.1.1 Tube Preparation. As with many other diffusion tube techniques, the sampler consists of a tube, two end caps and two grids. No membranes are used for NO_2 diffusion tubes. The tubes are of acrylic, and are 10.8 mm internal diameter, 71 mm long, with moulded or turned down ends to accept plastic caps. The two caps are of polyethylene, and are usually two distinctive colours. The grids are woven wire cloth, cut to a diameter sufficient to cover the end of the tube yet fit easily into the cap.

The grids, tubes and caps are cleaned in a detergent solution (which must be free of nitrogen compounds), for 30 minutes in an ultrasonic bath, and then rinsed with deionised water and dried in a drying cabinet. The grids are dipped in an absorbent mixture of triethanolamine and acetone, and placed on tissue paper to allow the acetone to evaporate. Two prepared grids are then placed in each cap of one colour which is fitted to the tube. Caps of a different colour are then used to close the other end of the tube.

These tubes are typically exposed for two to four weeks, simply by removing the cap which does not retain the grids. After exposure, the end cap is replaced to seal the tube until extraction and analysis.

3.1.2 Extraction and Analysis. The average NO_2 concentration for the exposure period may be determined by automatic and manual techniques. The automatic, and more common, technique is considered here. Samples which give concentrations that exceed the highest calibration solution may be diluted and the remaining solution reanalysed.

A known volume of demineralised water is added to the grids, and the tubes are gently swirled. The end cap is replaced and the tubes are allowed to stand for 30 minutes. To each tube in turn a sulfanilamide solution in orthophosphoric acid followed immediately by a NEDA solution is added. The caps are replaced, and the tubes are inverted, shaken, and left for 30 minutes to allow the azo dye to develop. The contents of each tube is then emptied into a disposable sample cup and place on an autosampler together with standards and drift standards.

Nitrite from the NO_2 collected reacts in the phosphoric acid solution with sulfanilamide giving a diazonium salt which couples with the naphthalene derivative to form an azo dye. The optical absorbance of the dye is measured at 540 nm, by an autosampler, which analyses a large number of samples. The amount of nitrite ion in the sample, and hence the mass of NO_2 collected, is obtained by reference to a calibration curve derived from the analysis of standard nitrite solutions.

3.2 Automatic Point Monitor Method for Oxides of Nitrogen

Automatic point monitors measure NO_x, $NO + NO_2$, using the gas phase chemiluminescent reaction of NO with O_3. Although issues regarding residence times under high concentrations have been raised in the past, these instruments are generally regarded as being very reliable if maintained properly. The measurement technique is well established, and instruments from differing manufacturers vary little.

The air sample is passed into the reaction chamber, where the NO reacts with an excess of O_3. This results in NO_2 in an excited state, which decays emitting a photon. The intensity of the chemiluminescent radiation is measured by a photomultiplier tube (PMT), the output voltage from the PMT being directly proportional to the NO concentration.

The incoming air sample is then redirected to pass over a heated molybdenum converter which reduces all of the NO_2 in the air sample to NO. This sample is then reacted with excess ozone in the same way as for the NO measurement, to give a total NO_x measurement. The NO_2 concentration is assumed to be the difference between the NO_x and NO measurements.

NO_x detectors may be modified to suit the specific use for which it is required. For example, roadside measurements will give high concentrations of both NO and NO_2 which may be prone to large fluctuations. It is thus possible that the residence time of the sample within the detector will not enable the detector to measure the true magnitude of the fluctuations of air concentrations. Under these circumstances it is appropriate to use a two channel instrument which measures NO and NO_x using the techniques outlines above, but with separate reaction chambers and detectors. Typical specifications for a single channel instrument are as follows (Bower *et al*, 1996):

Lower Detectable Limit		2 ppb
Precision as % of Upper Range Limit		<1%
Noise	Zero	1 ppb
	Span	3 ppb
24 Hour Drift	Zero	2 ppb
	Span	2% Full Scale
Linearity Error		2%
95% Response Time (max.)		180 seconds

Calibration of these detectors is usually conducted by using an internal source, as follows. Zero air is supplied to the reaction cell by passing ambient air through purafil and charcoal scrubbers. The span gas is generated by an NO_2 permeation tube, which contains a quantity of pure liquid NO_2. The permeation tube is kept in an oven at a constant temperature. Zero air is passed across the permeation source at a constant flow rate, and the amount of NO_2 permeating into the zero air stream gives a constant, and reproducible, NO_2 concentration. This gas is then passed in to the reaction cell and is used as the span gas. Calibration techniques which are traceable to a national standard are essential for quality control (see Chapter 8).

4 Sulfur Dioxide (SO_2)

4.1 Passive Diffusion Tubes Method for SO_2

4.1.1 Tube Preparation. The tubes and stainless steel grids are washed separately in an ultrasonic bath, with repeated rinses of deionised water. They are then dried overnight in an oven. No detergents are used, as most contain significant amounts of sulfur, either as an impurity or a constituent of the surface active molecule. The end caps are also cleaned in deionised water, dried, and stored in sealed plastic bags.

The tubes were prepared in a similar manner to O_3 diffusion tubes (section 2.1), with an adsorption solution of KOH. As no membranes are used, the end of the tube that does not have the grids is sealed until the tube is exposed. The preparation of the tubes is conducted quickly, to avoid contamination from the SO_2 in the atmosphere.

4.1.2 Extraction and Analysis. Hydrogen peroxide solution is added to the grids, and left to stand for at least one hour to complete the extraction and oxidation process. The SO_4^{2-} resulting from the oxidation of SO_2 is then analysed quantitatively by ion chromatography. A $NaHCO_3/Na_2CO_3$ solution is used as the eluant, to ensure good separation of the sulfate peak from the nitrate peak, which is evident in some samples. Calibration solutions are typically used every twelfth analysis to detect any drift in the response to the SO_4^{2-}. The SO_4^{2-} in each sample is calculated, and hence the SO_2 mass is determined. This is then corrected for reagent blank and travel blank as appropriate to give the SO_2 mass, which is converted into an SO_2 concentration during the exposure period of the tube.

4.2 Black Smoke and SO_2 Monitoring by Active/Semi-Automatic Samplers

The active technique employed for measuring black smoke is included in this section as it utilises the same equipment as the measurements for SO_2. Although automatic monitors are available for SO_2, the reduced costs associated with the active sampling technique are attractive, and it is thus possible to have a greater number of sites. In the UK, combined black smoke and SO_2 measurements have been made using active samplers for over 30 years. Consequently the measurement technique is not expected to be significantly modified in the future.

The following methods are used to determine daily average pollutant concentrations of these pollutants:

- SO_2:   Acid titration method, BS 1747 Part 3 (1991)
- Smoke: Reflectance method BS1747 Part 2 (1993)

Active samplers need to be located at fixed sampling locations with access to a power supply. These are usually in existing buildings or secure stand-alone site enclosures. In the well-known '8-port' sampler, outside air is drawn, via a funnel and length of tubing, through one of eight filter papers held in brass clamps. The particulate suspended in the air is deposited on the filter paper, producing a dark circular stain of known area. This filtered air sample is then bubbled through one of eight glass bottles containing acidified dilute hydrogen peroxide. The SO_2 in the air sample reacts with the solution to give sulfuric acid.

Tubes from the eight bottles are connected to an eight-port valve which changes every 24 hours, so that air is drawn through a fresh filter paper and bottle each day. The air sample is drawn through a gas meter before reaching the pump.

It is usual practice for the operator to visit every seven days. The operator takes a flow meter reading and collects the seven exposed filter papers and the seven exposed peroxide samples. Fresh filter papers are inserted into the clamps, and the bottles are refilled with fresh solution. The filter and bubbler currently active are left undisturbed, to be changed the following week.

The darkness of the smoke-stains are read using a reflectometer and expressed on a scale of 1 to 100. A standard calibration formula is used to convert this to a mass of 'black smoke' deposited. As the volume of air which has passed through the filter has been measured (typically 2 m^3), it is possible to calculate the concentration of particulate material, as 'black smoke', in the sampled air.

The bubbler solutions from the seven days are analysed for total acidity by titration with disodium tetraborate solution and, knowing the volume of sampled air, it is possible to calculate the concentration of total acidity expressed as SO_2. This assumes that the acidity in the solution is governed entirely by the presence of SO_2. The presence of other strong acids or alkali compounds in the air sample will affect the resulting acidity of the solution. Under most conditions, the impact on the pH from compounds other than SO_2 is negligible. However under conditions of high NH_3 concentration correction factors are required.

The methods outlined above for the measurement of SO_2 and black smoke are now standardised, and are contained in the following British Standard Specifications (and International Standards):

BS 1747 Part 2 1969 (1991) Determination of concentration of suspended matter
BS 1747 Part 3 1969 (1991) Determination of sulfur dioxide
BS 1747 Part 6 1983 (1990) = ISO 4219:1979 Sampling equipment used for the determination of gaseous sulfur compounds in ambient air.
BS 1747 Part 11 1993 = ISO 9835:1993 Determination of a black smoke index in ambient air.

4.3 Automatic Point Monitor Method for SO_2

SO_2 real time point monitors use a pulsed fluorescence technique. A UV pulsed light source is focused into the sample chamber, which elevates the SO_2 molecules into an excited state. The resulting radiation emitted from the decay of the excited state is measured by a photomultiplier tube (PMT). The output voltage from the PMT is proportional to the SO_2 concentration. A permeable membrane is used to remove interfering hydrocarbons before reaction. Typical specifications are listed below (Bower *et al*, 1996):

Lower Detectable Limit		2 ppb
Precision as % of Upper Range Limit		<1%
Noise	Zero	1 ppb
	Span	3 ppb
24 Hour Drift	Zero	2 ppb
	Span	2% Full Scale
Linearity Error		2%
95% Response Time (max.)		150 seconds

Calibration of the SO_2 detector uses a zero air supply and a span gas. The zero air is generated from ambient air by drawing it through a charcoal scrubber before entering the reaction cell. The span gas is generated by an SO_2 permeation source in a similar way to that for the NO_x analyser.

5 Carbon Monoxide (CO)

5.1 Automatic Point Monitor Method for CO

The automatic point analyser for CO uses an IR absorption technique. IR radiation at a wavelength of 4.5 to 4.9 nm is passed through the air sample, and the absorption is measured by an IR detector. A reference measurement is taken from an air sample with no CO present. It is thus possible to determine the absorption of IR that corresponds to the presence of CO, and hence the CO concentration in the air sample.

A similar method is also employed by other automatic point monitors, known as Non-Dispersive Infra-Red (NDIR). Here, differences in the IR absorption between the ambient air and the reference gas (containing no CO) cause a metallic membrane in the detector to move back and forth in accordance with the alternating gas flow and CO concentration.

Typical instrument specifications are as follows (Bower *et al*, 1996):-

Lower Detectable Limit		100 ppb
Precision as % of Upper Range Limit		<1%
Noise	Zero	50 ppb
	Span	200 ppb
24 Hour Drift	Zero	500 ppb
	Span	2% Full Scale
Linearity Error		2%
95% Response Time (max.)		60 seconds

Calibration of the CO monitor uses zero and span gases. The zero air is generated by passing ambient air through a heated palladium/alumina catalyst, before entering the reaction cell. The span gas is supplied from a dedicated CO cylinder, of known concentration.

6 Volatile Organic Compounds (VOCs)

6.1 Passive Diffusion Tubes Method for VOCs

Passive diffusion tubes give time averaged concentrations of benzene, toluene, ethylbenzene, *m+p*-xylene[1] and *o*-xylene (BTEX). Unlike previously mentioned passive samplers, the compounds that are held in the sorbent are removed by

---

[1] *m*-xylene and *p*-xylene elute at similar times, and are often not possible to resolve. Consequently they are, on occasion, expressed as a sum of the two components.

heating. Analysis is conducted using gas chromatography. The tubes are approximately 90 mm × 6 mm OD, and are of stainless steel. The diffusion path of the sampler is typically 10 mm, and the area of the sampler is approximately 30 mm^2. Stainless steel gauzes are used to retain the sorbent, and brass end-caps and unions with PTFE ferrules ensure air tight closure. Diffusive end-caps are fitted to the exposed end of the tube. These end caps are fitted with a gauze and an adjacent silicone membrane to allow the ingress of vapour but exclude water. There are several different sorbents which may be used for BTEX diffusion tubes. These are produced by different manufacturers and possess differing uptake rates for each of the compounds.

6.1.1 Tube Preparation.

6.1.1 Tube Preparation. Although preparation techniques and analysis parameters vary, the following is an example of how a tube is prepared, and then later analysed for BTEX.

Preparation of sorbent samplers: Prior to use, new tubes are washed in a detergent solution in an ultrasonic bath, and are rinsed well with distilled water. They are then dried in an oven, typically overnight. The brass end caps and unions are also washed as above. The gauzes are used as supplied. The sorbent packing is preconditioned by heating under an inert atmosphere, typically N_2. To prevent recontamination of the sorbent after preconditioning the tubes are kept in a clean atmosphere during cooling to room temperature, storage and loading into the tubes.

Gauze is inserted into one end of each tube. An appropriate amount of sorbent is then weighed, and poured into the end of the tube. The tube is gently tapped to ensure packing is evenly distributed. A second gauze is inserted, so that it presses onto the packing in the tube. A retaining spring is then located on top of the gauze. Both ends of the tube are tapped on a hard surface to check that no packing falls out. The pressure drop across each tube is checked and tubes which exceed 2 psi are rejected. Prior to use, the tubes are purged repeatedly at a high temperature, typically under helium, until the values for the components of interest do not exceed the blank values.

6.1.2 Analysis. Analysis of the BTEX tubes is conducted by gas chromatography. Precise methods vary, but the tubes are automatically sealed in line to an inert carrier gas stream, and are then heated, typically to approximately 200 °C, to release the VOC from the sorbent. The released compounds are passed through a cooled bed of sorbent, where they are trapped. After complete desorption from the diffusion tube, the cold trap is rapidly heated, and the released compounds are injected onto a capillary column in the gas chromatograph. As the compounds pass down the column, the temperature is increased. The column type and the temperature ramping are chosen to maximise compound separation. On eluting from the column, the compounds are detected by a flame ionisation detector (FID). Individual compounds are identified by the time taken to elute from the column. The resulting peak is integrated to generate an area, which is converted into a mass value, from the instrument calibration. By knowing the mass value, exposure time, and sorbent uptake rate for each compound, the average concentration of each of the compounds under investigation may be determined.

Calibration of the gas chromatograph is undertaken with either cylinders containing a known concentration of the relevant compounds, or injection of a known amount of the pure compound into the sampling system. The voltage output from the gas chromatograph for known amounts of compound is then used to generate a response factor. The frequency of calibration varies, typically response factors are determined at the start of each working day, with blanks and calibrations also being conducted during a batch analysis of diffusion tubes.

6.2 Active Diffusion Tubes for VOCs

It is possible to measure a greater variety of VOC with diffusion tubes by drawing air through the tube, and hence through the adsorbent bed. Flow rates for so called 'pumped diffusion tubes' vary typically from 10 to 200 ml minute^{-1}, with an exposure time of several hours the volume of air sampled depending on the airborne concentrations of VOC.

If the range of compounds to be measured and their concentrations are unknown, then three tubes, with different strength adsorbents may be connected together, so that the air passes through the weaker adsorbent first, and the strongest adsorbent last. These tubes are then analysed separately, and the most appropriate sampling method is determined from the results. For example, it may be sufficient to use passive diffusion tubes, or pumped diffusion tubes with more than one adsorbent may be required. If more than one adsorbent is appropriate, then these must be separated by glass wool, and the air sample must pass over the weaker adsorbent first.

The preparation and analysis of diffusion tubes with more than one adsorbent is similar to the method outlined for tubes containing single adsorbents, with the exception that the use of stronger adsorbents may require higher temperatures to desorb the compounds of interest.

6.3 Automatic Point Monitors for VOC and BTEX

There are two principal automatic detectors for measuring ambient concentrations of hydrocarbons. Both instruments use gas chromatography, with the more sophisticated analysers incorporating a cold trap to enable the trapping and measurement of a wide variety of VOCs, typically from the C_2 compounds to C_9. The cycle time of these detection systems is usually one hour. Detectors with no cold trap measure the less volatile compounds, and are tailored to measuring benzene, toluene, ethylbenzene, *p*-xylene, *m*-xylene and *o*-xylene (BTEX). These instruments typically have a cycle time of 30 minutes.

The principle of gas chromatography is outlined in the analysis of BTEX tubes (Section 6.1.2). The techniques used for sampling air directly differ slightly, in that the air sample is passed directly from the atmosphere to the cold trap, rather than originating from a diffusion tube. The cold traps are cooled to temperatures as low as $-90\,^\circ$C to facilitate trapping of the more volatile species. Depending on the instrumentation, this may require a cryogen, such as liquid nitrogen. The carrier gas is usually He, and FID detectors are used, which require a supply of clean air and hydrogen.

Instruments that are tailored to BTEX measurement use a similar trapping principle. Specific designs may vary. However no cryogen is required as the trap is not operated at reduced temperatures. A variety of carrier gases are used by different instruments. Some detectors use a precolumn, which enables a 'window' of compounds to be selected, and passed to the column for separation and detection. Other instruments use different techniques to accomplish the same result. As the volatility range of compounds to be identified is narrower, suitable separation of compounds (determined by the column type and the temperature cycle) may be achieved with a shorter cycle time. FID detectors are primarily used.

Analysis of the chromatograms is usually handled by tailor written software. Calibrations identify the elution time of each compound, and the peak area corresponding to a known concentration. A response factor, i.e. the peak area per ppb, for each compound is thus determined. This response factor is applied to the corresponding peak area measurements, and the concentrations are determined. Effective quality control is essential when generating the peak areas, and deriving the response factors from calibrations.

7 Lead and Heavy Metals

7.1 Lead (and Heavy Metal) Monitoring by Active Samplers

In the UK, ambient lead measurements are made in the lead network using a single port 'M-type' active sampler. Samples are collected on a filter paper (0.8 μm pore diameter membrane filter) by drawing air through the filter using a diaphragm pump. The volume of air is measured by a flow meter, located between the pump and the filter. The inlet is protected by a cylindrical hood, with the lower end open. This ensures protection from precipitation, but allows good air circulation. The sample flow is typically 4.5 l min^{-1} and is regulated by a critical orifice. Samples are changed weekly, although monthly intervals have also been used.

Samples may be analysed by different techniques. Currently Inductively Coupled Plasma Atomic Emission Spectrometry (ICP-AES) and Inductively Coupled Plasma Mass Spectrometry (ICP-MS) are the two most common methods. ICP-AES uses an argon plasma to excite the atoms, and light of characteristic wavelength is emitted as the atoms return to the ground state. Quantification is obtained by comparing the output with samples of known concentration. This analysis technique can measure 45 elements simultaneously, with varying detection limits. ICP-MS uses an argon plasma as an atmospheric pressure ionisation source for a quadrupole mass spectrometer. In general, this technique gives greater sensitivity than ICP-AES, and more elements may be quantified.

7.2 Mercury Vapour (Hg) Automatic Point Monitor

The instrument operates by trapping the Hg vapour from the air in an ultra pure gold adsorbent. The amalgamated Hg is then thermally desorbed, and detected using Cold Vapour Atomic Fluorescence Spectrometry (CVAFS). The air sample is passed into one of two cartridges, where the Hg is adsorbed in the gold matrix. The switching between two cartridges allows measurements to be made from one, whilst the other is sampling. Hg is then released by heating the gold in ultra pure argon. The Hg is carried into a quartz cuvette illuminated by a low pressure Hg vapour lamp. This excites the Hg vapour atoms, which emit at 253.7 nm. The emitted radiation is detected by a photomultiplier tube, with the intensity being directly proportional to the Hg vapour concentration. The interior of the detector is purged with argon carrier gas, to eliminate possible interference caused by ozone generation in the optical path.

Calibration may be conducted using a permeation source, or a manual injection of known concentration. Inert gas is used to provide a continuous purge flow for the permeation source, and this ensures that no oxidation occurs on the surfaces of the permeation tube.

8 Particulate Matter and Black Smoke

Particulate matter is described using different terms, and it is important to clarify the precise measurement that is being made. One technique for measuring black smoke utilises the same equipment that is used for measuring SO_2, and consequently the technique is discussed in detail in Section 4.2.

8.1 Active Total Particulate Samplers

There are several types of instrument which collect particles by drawing the air sample through a section of filter paper for a specified time. At the end of the exposure period the roll of filter paper is wound on, so that a clean section of paper is presented to the air stream. Analysis for total particulate is relatively simple, with a known area of the sample being removed and weighed. As the weight of the filter paper is known, the excess mass is attributed to the collected particles. Chemical analysis of the samples may also be conducted by ion chromatography to determine the concentrations of specific ions.

Similar weighing techniques may be employed with size fractionated measurements. Impactors separate the particles into size fractions by causing the air stream to be diverted through a series of staggered slits. Particles that are too heavy to be deflected pass through a slit and impact on a filter paper. As the deflections become successively more severe, the lighter particles are captured, until a back filter captures any particles remaining in the air stream. The weight increase of the filter papers are attributed to the accumulation of particulate matter. These techniques are typically used to average over long periods of time, to decrease the uncertainties in the weighing of the filter papers.

8.2 Automatic Point Monitors Methods for Particulate Material

8.2.1 Tapered Element Oscillating Microbalance (TEOM). Particles of aerodynamic diameter less than 10 μm are of special importance because of their possible detrimental health effects and are known as PM_{10}. One automatic instrument for determining PM_{10} is known as the tapered element oscillating microbalance (TEOM). This technique operates by continuously measuring the weight of particles deposited onto a filter. The detector head has an aerodynamic cut-off of 10 μm, and consequently the detector measures PM_{10} rather than total particulate. The filter is attached to a hollow tapered element which vibrates at its natural frequency of oscillation. As particles progressively collect on the filter, the frequency changes by an amount proportional to the mass deposited. As the air flow through the system is regulated, it is possible to determine the concentration of PM_{10} in the air. The filter requires changing periodically, typically every 2 to 4 weeks, and the instrument is cleaned whenever the filter is changed.

8.2.2 Beta Attenuation. Beta attenuation monitors operate by collecting particulates on a filter paper over a specific cycle time. The attenuation of beta particles through the filter is continuously measured over the cycle time. This gives a real time measurement of either total particulate, PM_{10} or $PM_{2.5}$, depending on the configuration of the instrument.

At the start of the cycle, air is drawn through a glass fibre filter tape, where the particulates deposit. Beta particles that are emitted from either a C_{14} or a K_{85} source are attenuated by the particles collecting on the filter. The radiation passing through the tape is detected by a scintillator and photomultiplier assembly. A reference measurement is made through a clean portion of the filter, either during or prior to the accumulation of the particles. This reference measurement enables baseline shifts to be corrected for.

Different inlet arrangements are used to configure the instrument to measure either total suspended particles, PM_{10} or $PM_{2.5}$. The sample inlet may incorporate a heater to eliminate problems associated with varying humidity.

9 Toxic Organic Micropollutants (TOMPS)

The following compounds are collectively known as Toxic Organic Micro-pollutants (TOMPs): polychlorinated dibenzo-*p*-dioxins (PCDDs), polychlori-nated dibenzofurans (PCDFs) polyaromatic hydrocarbons (PAHs) and polychlorinated biphenyls (PCBs). These compounds have significant vapour pressures at ambient temperatures. Hence when sampling these compounds sample of both the gas and the particles present must be taken. A filter is used to trap the particle phase component, and the gas phase component is trapped with an adsorbent e.g. precleaned polyurethane foam (PUF) or XAD-2 resin. Different sampling techniques are used for ambient and stack monitoring.

9.1 The PUF Sampler

The ambient atmospheric sampler (known as the PUF sampler) comprises of a standard pesticide sampler fitted with a rectangular head. The head is modified so that the sampler operates in a non-directional manner, and is less restricted with respect to the upper particle size capture. Air is drawn, by a fan, through the circular annulus of the sampling head, through a filter paper, and the polyurethane foam PUF plugs. The time which the sampler operates is recorded by an hour meter and the flow rate is determined using a calibrated orifice. Each sampler is fitted with a pressure transducer and a data logger which records the pressure drop throughout the sampling period so that the sampling rate can be accurately determined.

It is also possible to investigate the deposition rates of PCDD/Fs, PAHs and PCBs. This is done by exposing a PTFE coated deposition gauge, and analysing the samples using the procedure as outlined below.

9.2 Sample Preparation, Handling and Extraction of TOMPS

Quality control in relation to sampling and analytical procedures is extremely important. In order to ensure the quality of the results, ^{13}C labelled and other standards may be introduced at various stages in the procedure and the recovery of these standards determined the acceptability of the results. Standards for PCBs, PAHs and PCDD/F may be generated from the pure compounds.

The polyurethane plugs used in the PUF samplers can become contaminated with a wide range of compounds, including PCDD/Fs and PCBs. Any potential contaminants are removed from the plugs by a lengthy cleaning process. The cleaned PUFs are individually wrapped in aluminium foil and stored in sealed polythene bags until required. Isotope labelled tracers are added to the filter holder and PUFs before they are used. This is in order to identify and quantify losses of compounds from the operational exposure of the PUFs.

Following exposure, the sampler cartridges are dismantled in a clean air cabinet. The PUFs and filters are bagged and stored in a refrigerator at 4 °C prior to analysis. The basis of the analysis procedure is to add analytical recovery standards and then to extract both the standards and material captured on PUFs and filters into a suitable solvent. Hexane is used as a suitable solvent for PAHs and PCBs and toluene for PCDD/Fs.

9.3 Analytical Methods for TOMPS

Different compounds are suited to the different analysis techniques, as given below. Analytical isotope labelled standards are introduced at the start of the procedure to act as internal standards.

9.3.1 PCDD/Fs.   The state of the art analysis of PCDD/Fs is carried out by a combination of gas chromatography and mass spectrometry. The PCDD/Fs are extracted from the sample in toluene overnight. Isotope labelled materials are added before extraction. The volume of the extract is reduced and it is then purified to remove any contaminants which may interfere with the later quantification. The cleaned extract is injected into a gas chromatograph. This separates the 17 PCDD/F of interest from the other compounds present. The mass spectrometer is used to quantify the results by comparing the quantity of the isotope labelled extraction spike with the sample.

9.3.2 PCBs.   PCB analysis is carried out by several approaches depending on the sample. The analysis of stack samples is generally carried out following a similar approach as is used for PCDD/Fs. However, because PCBs are often found at concentrations 2 or 3 orders of magnitude higher than the PCDD/Fs an electron capture detector (ECD) is sometimes used. In the case of GC-ECD, quantification of the different analytes is on the basis of peak area and the use of internal standards.

9.3.3 PAHs.   Some PAHs are of higher volatility than the PCDD/Fs. Hence a solvent with a lower boiling point (such as hexane) is used for their extraction to reduce volatilisation losses. The purification approach used also cannot be as vigorous without leading to unacceptable losses of the compounds of interest. This is measured by deuterated PAHs added before extraction.

Two main quantification routes are used for PAHs in the UK. These are high performance liquid chromatography (HPLC) followed by fluorescence detection, or gas chromatography followed by mass spectrometry. The former approach has the advantage of being less expensive but the latter provides greater specificity. Results from the TOMPS network may be found in Branson *et al*, 1997.

10 Ammonia (NH₃)

Diffusion tubes, which use the same principles outlined in Sections 2.1, 3.1 and 4.1, may be used for the determination of ammonia. The diffusion tubes are coated with oxalic acid, and then analysed, often using flow injection analysis with conductometric detection.

10.1 Active Filter Pack Samplers for NH₃

The filter pack sampler consists of a PTFE prefilter (1 μm pore size) to capture particulate NH_4^+. This is followed by an acidified Whatman 42 filter paper to extract the NH_3 from the air drawn through the filter pack (Pio, 1992). The filter paper is impregnated in H_3PO_4 and dried in clean air to reduce the levels of NH_3 before exposure in the field. Sample flow rates are typically 10 to 12 litres min^{-1} and are measured by an attached gas meter. A 2 hour exposure is usually used, although this is dependent on the expected NH_3 concentrations. Exposed samples, and blanks, are stored in clean sealed tubes until analysis. After extraction on a propan-2-ol solution, the samples are analysed using an indophenol blue reaction.

10.2 Active Denuder Samplers for NH₃

Denuders are hollow glass rods, which have the internal surface coated with an adsorbent coating - usually oxalic acid, although citric acid is also used (Pio, 1992). Air is then drawn through the denuder at a known rate, typically regulated by a critical orifice. The NH_3 in the air diffuses across to the surface, where it is adsorbed. Denuders typically have an exposure time of several hours, depending on the expected airborne concentrations of NH_3. As small section of the denuder is not coated, so that air passing into the denuder establishes laminar flow prior to reaching the coated surface. This ensures that the adsorption to the internal surface is by diffusion alone. A direct consequence of this is that the length of the denuder is determined by the diffusion coefficient of NH_3, the internal diameter of the denuder and the flow rate. The samples are then extracted by rinsing the denuders with a known volume of demineralised water. These extracts are then analysed for NH_4^+ by using ion chromatography.

10.3 Rotating Wet Denuder - Active and Automatic - for NH₃

There are two NH_3 detectors which operate on a similar principle, one of which is an active sampler, and the other is an automatic monitor. Both of the instruments use a rotating annular denuder, which is part filled with a stripping solution of sodium bisulfate. Rotation of the denuder ensures that the surfaces are coated with stripper solution, which facilitates efficient capture of NH_3. The air sample is continuously drawn through the denuder, and the NH_3 is adsorbed into the solution.

The active sampler drains the denuder at pre-programmed intervals, typically hourly, and stores the solution in a clean container located on a carousel. The denuder is then refilled with fresh stripping solution, and the carousel is rotated to allow the next sample to be stored. The samples are sealed and removed periodically. Analysis of the samples may be conducted by a variety of techniques.

The automatic detector continuously drains the stripping solution from one end, and replenishes the rotating denuder with fresh solution at the other end of the denuder. The removed stripper solution is passed through a debubbler and mixed with NaOH to convert the NH_4^+ to gaseous NH_3. The solution is passed along a PTFE membrane, which is permeable to gases, but not liquids. Some of the NH_3 diffuses through the membrane into a stream of pure water. The water is purified by using a membrane to remove bubbles, and an ion exchange column to remove pollutants, and hydroxyl ions are removed by using an NH_4^+ solution in NaOH. The sample solution then flows through the conductance cell, immediately followed by a temperature sensor. The conductance is converted into concentration values through information obtained during calibration. The detector is calibrated using standard solutions and a blank. The calibration solutions are sampled in the same way as the air with exception that the denuder is bypassed. The response of the conductance cell, to the calibration solutions is determined so that values may be converted to concentrations.

11 Acid Deposition

11.1 Acid wet-deposition monitoring by active sampler methods

Wet deposition samplers are used to provide information on wet deposition of ionic species such as sulfate, nitrate and ammonium as well as acidity. The collectors consist of a funnel covered by a firmly fitting lid which is opened automatically during rainfall. Precipitation is stored in the collection vessel at a temperature of 4 °C to prevent sample degradation, and an automatic logging facility records the time of onset and cessation of precipitation. Up to 8 daily samples can be collected automatically, after which the samples are analysed for conductivity and pH. The concentrations of the cations NH_4^+, Na^+, K^+, Ca^{2+}, Mg^{2+} and the anions NO_3^-, Cl^-, SO_4^{2-} and PO_4^{3-} are determined using IC. In addition to this automatic wet-only instrument, samples may also be taken of total deposition in simple collectors. These are then analysed using the same techniques.

12 Hydrogen Peroxide (H_2O_2) and Organic Peroxides

H_2O_2 may be measured by the chemiluminescent reaction between H_2O_2 and luminol, and a fluorescence technique using the liquid phase reaction of peroxides with *p*-hydroxyphenylacetic acid.

12.1 Chemiluminescence Technique for Measuring Peroxides

The chemiluminescence technique passes the air sample through a coil with a stripping solution of NaOH and $NaHCO_3$. The H_2O_2 in the air diffuses into the solution, which is then passed through a phase separator to remove the gas phase. The solution is passed to the reaction chamber where it is mixed with a luminol solution. The light emitted from the resulting reaction is detected by a photomultiplier tube, with the resulting voltage output being proportional to the H_2O_2 concentration in the sample solution, and hence the airborne concentration (Davies and Dollard 1989). As H_2O_2 degrades over time, solutions that are used for calibration purposes first require titration to determine the amount of H_2O_2 present. H_2O_2 reacts with potassium permanganate when acidified with dilute sulfuric acid, which proves to be convenient reaction for titration, as the potassium permanganate can be used as the indicator.

12.2 Fluorescence Technique for Measuring Peroxides

The fluorescence technique uses the liquid phase reaction of peroxides with *p*-hydroxyphenylacetic acid, catalysed by peroxidase. This reaction produces a fluorescent dimer. The instrumentation is similar to that for the chemiluminescence technique, comprising of a stripper coil, phase separator and a detection cell. Reagents are mixed prior to entering the fluorescence flow cell.

Two channels are used to distinguish between H_2O_2 and organic peroxides. For measurement of organic peroxides, the H_2O_2 is selectively destroyed by catalase prior to detection. The amount of H_2O_2 is then given by the difference between total peroxides and organic peroxides. The exact methodology is complex, involving corrections for the catalase destruction efficiency, and several other parameters. The peroxides in the fluorescence flow cell are excited by a Cd lamp at 326 nm, and the emitted radiation, between 400 and 420 nm, is detected by photomultiplier tubes. The output is corrected for fluctuations in the intensity of the Cd lamp. H_2O_2 and the organic peroxides may also be separated by HPLC followed by derivitization and fluorescence detection.

13 Nitric Acid (HNO₃)

Nitric acid (HNO_3) concentrations can be measured using denuders. The denuders comprise a length of straight glass tubing, of approximately 1 metre, with the internal surface coated in NaF. The denuders are mounted vertically, and air is drawn through the denuders at a rate of approximately 2 l min^{-1}. The flow through the denuders is measured by a flowmeter, and controlled by a critical orifice. The HNO_3 from the air reacts with a NaF coating on the inner surface of the denuder, forming $NaNO_3$. A small section of the internal surface is uncoated, and this ensures that laminar flow is established, before the air passes over the NaF. Nitrate (NO_3^-) from ammonium nitrate is assumed not to deposit to the walls. After exposure for several hours, the denuders are removed, and the ends are sealed until extraction and analysis. The $NaNO_3$ is extracted by drawing 10 ml of analytical grade deionised water up through the internal coated surface of the denuder several times. These samples are then analysed for NO_3^- using ion chromatography.

14 Other Measurement Techniques

There are a number of long pathlength measurement techniques which are employed to measure a variety of pollutants. Although expensive, these techniques are often flexible, and enable the simultaneous measurement of several different compounds. Optical remote sensors can be grouped into two different classes: monochromatic (laser long path absorption, Differential Absorption Lidar) and spectrally broad band (Fourier Transform IR Spectrometer, UV Spectrometer). Examples of each are given in the following sections, and more detailed information may be found in Grant *et al*, 1992.

14.1 Fourier Transform IR Spectroscopy (FTIR)

Most IR remote sensing techniques are suited to the measurement of single compounds. However FTIR has the advantage of measuring many compounds without scanning across a spectral range. As a result it is much more efficient at collecting, and spectrally analysing radiation.

An IR source is located at the far end of the path, with a receiver that focuses the incoming radiation. A scanning Michelson interferometer is used to split and recombine the incoming radiation. This causes interference between the two beams, arising from the phase difference, which is dependent on the wavelengths present. The radiation in one of the paths is reflected off a moving mirror, resulting in an intensity variation which is measured at the detector as a function of path length difference. This results in an interferogram.

The broadband interferogram is a sum of cosine waves for each spectral component as a function of pathlength difference. It is therefore possible to perform a Fourier transform on the interferogram, to obtain a spectrum in optical frequency units (cm^{-1}). Standard analysis packages are commercially available to process the interferogram, and produce concentration values. Under certain conditions it is possible to use the IR emissions spectrum of compounds to determine their concentrations.

14.2 Differential Optical Absorption Spectroscopy (DOAS)

While the vast majority of molecules adsorb in the IR spectral region, there are several that overlap with the adsorption of H_2O or CO_2, making measurement extremely difficult. Additionally, some compounds have no IR spectra. Consequently, the use of UV spectral techniques act as alternatives.

DOAS is typically used in the UV and visible spectral regions, being an adsorption technique. The source is a remotely positioned UV-visible light source, such as a xenon arc lamp. It is also possible to use a co-located lamp with a retro-reflector to increase the pathlength. The incoming radiation is reflected into a diffraction grating, to facilitate spectral separation, and measured by a photodiode detector array. There are several different techniques employed to scan the frequency range, the most common using diffraction grating to select the frequency. This is logged in conjunction with the output from the spectrometer, to give the resulting absorption spectrum.

14.3 Laser Long-Path Absorption

Lasers can be used to supply a collimated monochromatic beam of radiation that in many cases can be tuned to an absorption feature of the species of interest. This beam can be directed to a retroreflector (or even a topographic target), for long-path measurements. Additionally it is possible to draw air into a sample tube; the laser path is multiply reflected along the length of the tube to increase the absorption.

Differential absorptions are measured using lasers tuned to discrete lines. The path lengths used for this technique vary, depending on the power of the laser, and whether pulsed or continuous wave laser systems are used. The absorption information gives concentrations averaged across the path length. However, by taking measurements at different angles and ranges, it is possible to obtain range resolved information. Lasers used for this technique are primarily tuneable diode lasers (although the CO_2 laser is also used). Clearly, the laser source is required to be very stable.

14.4 Differential Absorption Lidar (DIAL)

The DIAL instrumentation uses two pulsed laser sources, one of which is tuned to the absorption frequency of interest, the other tuned to a slightly different frequency. The backscattered radiation is measured by a suitable detector co-located with the sources. Use of the two sources enables the differential absorption to be determined. Backscattering (Rayleigh and Mie scattering) of other atmospheric components can be accounted for. Range resolved information may be obtained from the time taken for the back-scattered signal to return to the source/detector. Consequently this technique is particularly of use when investigating dispersion of pollutants.

15 Future Developments

The increasing demand for air quality measurements and for improvements to their sensitivity and spatial and temporal resolution will continue to drive research into new techniques.

Diffusion tube preparation and analysis may undergo some refinements, and expand to incorporate several other compounds; however the basic technique is not expected to change significantly. Similarly, with several of the automatic point monitors, the instrumentation is so well established that no significant changes are expected in the near future. Clearly this does not preclude new measurement techniques becoming equally attractive.

As a general trend, it is expected that techniques currently used for short term field projects will become cheaper, and may be specifically tailored to long term monitoring. The development of new measurement techniques is governed by demand. This in turn is linked to any new legislation relating to air quality. Consequently it is possible that changes in the focus of air monitoring will lead to new techniques becoming commercially available. An example of is the measurement of VOC. The instrumentation required to measure a wide range of VOC is expensive, and currently the primary interest focuses on 1,3-butadiene and benzene alone. It is expected that automatic BTX instruments (as detailed in Section 6.3), will be modified to incorporate 1,3-butadiene measurement within the next several years.

There are other measurement techniques of note that are in the development stages. Gas chromatography is a versatile measurement technique, and the use of different trapping arrangements and columns enables the detection of many different compounds. One of the main problems associated with the technique is ensuring that reactive species are trapped and injected onto the column without any losses. This is an area of ongoing research and improvement, which is expected to result in an increasing number of compounds being measured by this technique.

The use of solid state detectors is also being developed, and it is expected that this new technique will make a considerable contribution to air quality measurement in the near future. Several investigations have been conducted, both in the field and the laboratory, using established techniques for comparison.

16 National Monitoring Networks in the UK

The Department of the Environment funds a number of national-scale air quality monitoring programmes throughout the UK. These are organised into 3 automatic networks and 6 sampler-based programmes measuring a comprehensive range of pollutants including sulfur dioxide, particulate matter, nitrogen oxides, ozone, carbon monoxide, hydrocarbons, lead, acid deposition and air toxics. There are currently over 60 automatic point monitoring stations and more than 1200 sampler measurement sites throughout the UK.

Different measurement methods are used to meet the different needs of each network (see Table 2). Complex real-time automatic measurements are used to acquire high resolution data, whilst simple and cost-effective sampler methods are used for surveys of air quality covering large areas of the UK. Site locational criteria, operational practice and QA/QC structures may also differ from network to network; their common aim is, however, to provide reliable and high quality measurements of air quality throughout the UK. Information from the networks are made readily available through several channels. Data from several of the networks is located on the National Air Quality Information Archive (Internet Address: http://www.aeat.co.uk/netcen/aqarchive/archome.html), and is published in annual *Air Pollution in the UK* reports (Broughton *et al*, 1997).

The objectives of the national air quality monitoring programmes are:

- to understand air quality problems so that cost-effective policies and solutions can be developed;
- to assess how far standards and targets are being achieved;
- to provide public information on current and forecast air quality;
- to assist the assessment of personal exposure to air pollutants.

16.1 Automatic Urban Network

Following the publication of the 1990 UK Government White Paper 'This Common Inheritance' the Department of the Environment commenced its Enhanced Urban Monitoring Initiative. This involved establishing an enhanced urban monitoring network (the EUN) with multi-pollutant monitoring sites in major cities throughout the UK. The current Automatic Urban Network (AUN) was later formed by the amalgamation of the former Statutory Urban Network (SUN) and the Enhanced Urban Network (EUN).

The SUN was originally designed and established in 1987 to monitor for compliance of ambient air quality with EC Directive legislative standards for O_3, NO_2 and SO_2. The EUN was primarily intended to provide up-to-date air quality information to the public. The new AUN meets both of these objectives, as well as providing a comprehensive source of information for policy and research purposes. More recently, many local authority-owned monitoring sites have also been integrated into the AUN. Plans for future network expansion are expected to result in an increasing number of sites.

The network uses fixed point real-time continuous analysers to monitor ambient levels of five principal air pollutants on an hourly basis:

- sulfur dioxide (SO_2)
- oxides of nitrogen (NO & NO_2 or NO_x)
- ozone (O_3)
- carbon monoxide (CO)
- particulates (PM_{10})

The stations are situated in a variety of locations. Many are in 'urban background' locations, to represent levels of pollutants to which large numbers of people may be exposed; these include sites such as city centre shopping precincts and central squares. However, in order to obtain a more complete picture of the concentrations to which people are exposed in different environments, stations are also situated in a variety of kerbside, suburban and industrial locations.

16.2 Automatic Rural Network

The objectives of this network are to provide information on rural pollutants throughout the UK. The network objectives are to enable environmental and health impacts to be assessed, to provide data required by the EC Directive on ozone, to enable policy development and provide public information. All sites measure ozone in rural areas, whilst some also measure SO_2 and NO_x in order to provide additional information to help understand the nature of photo-chemical episodes.

16.3 The Automatic Hydrocarbon Network

This network constituted the second phase of DoE's Enhanced Urban Monitoring Initiative, and measures a range of hydrocarbon species in urban air on a continuous basis (Dollard *et al*, 1995). These species have been selected either for their potential in producing photochemical oxidants such as ozone, or for their likely health impacts on the public at large. Information on two potentially toxic compounds, benzene and 1,3-butadiene, is rapidly broadcast to the public. In addition to the instruments on the network, VOC and BTEX detection systems are currently being affiliated into the network. The 26 compounds[1] measured by the instruments on the network are:

ethane	*cis*-2-butene	*n*-hexane
ethene	iso-pentane	benzene
propane	*n*-pentane	*n*-heptane
propene	1,3-butadiene	toluene
iso-butane	*trans*-2-pentene	ethylbenzene
n-butane	*cis*-2-pentene	(*m+p*)-xylene
ethyne	2-methylpentane	*o*-xylene
trans-2-butene	3-methylpentane	1,3,5-trimethylbenzene[1]
1-butene	isoprene	1,2,4-trimethylbenzene[1]

Due to the complexity of the instrumentation, and the number of compounds that are measured each hour, it is essential to have effective QA/QC procedures in place. Results are published in the Air Pollution in the UK reports, and is placed on the National Air Quality Information Archive.

16.4 NO$_2$ Diffusion Tube Network

In order to improve the spatial coverage of national NO_2 measurements, and to monitor NO_2 levels over the period of the introduction of catalytic converters on cars, a large long-term diffusion tube survey was initiated in 1993. Currently there are over 1200 measurement sites in operation, with 295 local authorities taking part in the network. Each local authority operates at least four diffusion tube sites: one kerbside, one intermediate and two in background locations. Monthly average NO_2 concentrations are made available on an annual basis (Stevenson K.J., Bush T.J. and Mooney D.E., 1997).

---

[1] The C_9 compounds are not routinely reported from the network instruments, however some of the affiliated sites are capable of measuring these compounds.

16.5 Black Smoke and SO_2 Network

These networks have now been operating for over 30 years and have success-fully monitored the massive decline in the concentrations of these pollutants resulting from the implementation of the Clean Air Acts. The 252 sites are distributed amongst two networks:

Basic Urban Network. This is a network of stations located in a representa-tive range of urban locations throughout the UK, established to provide information on long-term trends and to calculate national average concentra-tions on a consistent basis.

EC Directive Network. A network of stations in areas of relatively high concentration, targeted to monitor for compliance with the EC Directive on Smoke and SO_2. Daily average concentrations from all stations are published annually (Stevenson *et al*, 1997).

16.6 Multi-element and Lead Network

The Multi-element and Lead Networks provide long-term concentration measurements for 10 elements (Cd, Co, Cr, Cu, Fe, Mn, Ni, Pb, V and Zn). The measurement techniques employed are outlined in Section 7.1.

16.7 Acid Deposition Network

The Acid Deposition Network consists of 32 sites. Samples of wet deposition are collected and analysed for conductivity, pH, and various cations and anions.

16.8 Rural SO_2 Network

The Rural SO_2 Network provides daily measurements of rural SO_2 at 29 sites. The sampling methodology is outlined in Section 4.2.

16.9 Toxic Organic Micropollutants (TOMPS) 'Network'

To meet its national and international policy objectives, the UK Department of the Environment supports the operation of a monitoring network to determine environmental levels of TOMPS. These include PCDD's, PCDF's, PAH's and PCB's. The deposition and concentrations of Toxic Organic Micropollutants (TOMPS) in the atmosphere are measured at four sites. Results are published in Annual Reports (Branson *et al*, 1997).

Table 2   *UK national air quality monitoring networks (1/8/96)*

Network	Urban	Hydro-carbons	Rural	Diffusion tube	Smoke/ SO_2	Lead + Elements	Acid Deposition	Rural SO_2	TOMPs
Pollutants	O_3 NO_x SO_2 CO PM_{10}	25 Species	O_3 NO_x SO_2	NO_2	Smoke SO_2	Pb Metals	Anions and Cations	SO_2	PAHs PCBs Dioxins
Site Numbers	46**	12***	16*	1190	222	25	32	29	4
Measurement Techniques	A	A	A	PS	AS	AS	AS, PS	AS	AS
Function	S	N	S	N	S	S	S	N	N

A = Automatic   AS = Active Sampler   PS = Passive Sampler
S = Statutory   N = Non-statutory

*   Includes one site operated by PowerGen
**   Includes local authority aliated sites
***   Hydrocarbon monitoring at Birmingham East, Middlesborough and Southampton Centre is co-located with the Urban site

Acknowledgements

The authors wish to acknowledge the following departmental members for their contributions: B. Stacey, T. Bush, K. Stevenson, A. Loader, D. Mooney, B. Jones and Peter Coleman.

References

Many of the instrumentational details have been taken from manufacturers literature, such as instrument manuals. It is not considered appropriate to list these sources here.

1. Bertorelli V. and Derwent R. (1995). Air Quality A to Z. Compiled by the Meteorological Office for the Department of the Environment.
2. Bower J.S., Vallance-Plews J., McGinlay J., Stacey B.R.W., Telling S.P., Christiansen S.B., Eaton S.W., Broughton G.F.J., Willis P.G., Stevenson K.J. and Charlton A.J. (1996). Automatic Urban Monitoring Network - Site Operators Manual. Compiled by AEA Technology for the Department of the Environment.
3. Bower J.S. (1997). Ambient Air Quality Monitoring. A Review Paper for the Royal Society of Chemistry.
4. Branson J.R., Coleman P.J., Donovan B.J., Nicholson K.W., Watterson J.D., Jones K., Halsall C. and Lee R. (1997). Results from the Toxic Organic Micropollutants (TOMPS) network: 1991 to 1995. AEAT/16419146/REMA-227, A report produced for the Department of the Environment.
5. Broughton G.F.J., Bower J.S., Willis P.G. and Clark H.P. (1997). Air Pollution in the UK: 1995. Compiled by AEA Technology for the Department of the Environment.

6. BS 1747 Part 2 1969 (1991) Determination of concentration of suspended matter.

7. BS 1747 Part 3 1969 (1991) Determination of sulfur dioxide.

8. BS 1747 Part 6 1983 (1990) = ISO 4219:1979 Sampling equipment used for the determination of gaseous sulfur compounds in ambient air.

9. BS 1747 Part 11 1993 = ISO 9835:1993 Determination of a black smoke index in ambient air.

10. Davies T.J. and Dollard G.J. (1989). Measurements of gaseous Hydrogen Peroxide in Rural Southern England. AERE R13095 A report produced for the Department of the Environment.

11. Dollard G.J., Davies T.J., Jones B.M.R., Nason P.D., Chandler J.E., Dumitrean P.J., Delaney M., Watkins D. and Field R.A. (1995). The UK Hydrocarbon Monitoring Network. Volatile Organic Compounds in the Atmosphere, ed. Hester and Harrison, Issues in Environmental Science and Technology, 4, The Royal Society of Chemistry pp 37-50.

12. Grant W.B., Kagann R.H. and McClenny W.A. (1992). Optical Remote Measurement of Toxic Gases. Journal of the Air and Waste Management Association Vol. 42 No. 1 pp 18-30.

13. McGinlay J., Vallance-Plews J. and Bower J.S. (1996). Air Quality Monitoring: A Handbook for Local Authorities. AEA/RAMP/20029001. Prepared by AEA Technology for the Department of the Environment.

14. National Air Quality Information Archive (1997). Internet Address http://www.aeat.co.uk/netcen/aqarchive/archome.html Operated by AEA Technology for the Department of the Environment.

15. Pio C. Measurement of NH_3 and NH_4^+ in the Atmosphere by Denuder and Filter Pack Methods. Evaluation of the Rome Field Intercomparison Exercise. Development of Analytical Techniques for Atmospheric Pollutants, CEC and CNR Workshop, April 1992 pp 239-252.

16. Stevenson K.J., Bush T.J. and Mooney D.E. (1997). UK NO_2 Diffusion Tube Survey, 1995. Prepared by AEA Technology for the Department of the Environment.

17. Stevenson K.J., Loader A.E., Mooney D.E. and Lucas R. (1997). UK Smoke and SO_2 Monitoring Networks 1995-1996. Prepared by AEA Technology for the Department of the Environment.

Quality Assurance and Quality Control of Ambient Air Quality Measurements

B. Sweeney, P.G. Quincey, M.J.T. Milton and P.T. Woods

Environmental Standards Section, National Physical Laboratory, Teddington, Middlesex, TW11 0LW, UK

1 Introduction

Comprehensive measurements of ambient air quality require the commitment of significant resources, which can only be justified if the results are widely used and recognised as authoritative. This leads to a specific requirement that they should be accurate, traceable and representative. This in turn leads to the requirement for the use of effective quality assurance and quality control (QA/QC) procedures. In most cases, these QA/QC procedures will only account for a small fraction of the total cost of the measurements, but they will ensure that the measurements fulfill all their objectives and can be widely-recognised as authoritative.

The main issue involved in the QA/QC of ambient air quality measurements are described in the following sections.

2 Design of an Air Quality Monitoring Programme or Network

It is essential that the overall aims and objectives of the monitoring programme are defined prior to any sites being selected or equipment being procured. For example, different measurement scenarios reflect different interests in ambient air quality. Measurements designed to monitor long-term trends in ground-level ozone concentrations on a national scale will require a different set of monitoring requirements, when compared to measurements aimed at quantifying the effects of altering traffic flows in an urban environment.

After the objectives of the study are defined, the following issues should be considered:

2.1 Optimum Number of Monitoring Sites

If urban networks are needed in a densely populated major conurbation with a multiplicity of differing sources, it is useful to have analysers located relatively close to each other on a grid system. However, measurements at rural locations may not require such a system, although there is a different requirement to select representative sites to cover broad geographical areas.

2.2 Type of Sampling Methodology

The overall aims of the programme, and the resources allocated to it, will define the sampling methodology to be used. For instance, in assessing sulfur dioxide concentrations due to plume grounding around a power station, inexpensive passive samplers such as diffusion tubes may be employed to produce monthly-averaged concentrations. However, if long-term average concentrations are not relevant (for example if the measurements are to assess compliance with EC Directives, which require hourly values), then continuous analysers will need to be employed.

In certain circumstances it may be advantageous to employ both continuous and passive/active samplers during specific monitoring programmes. The use of continuous analyser(s) will allow high temporal resolution measurements to be made at particular locations, whilst the passive samplers will cost-effectively provide coverage over a wider geographical scale.

2.3 Measurement Requirements

This issue, which covers the accuracy and the spatial and temporal resolution of the measurements, will also have a bearing on the type of instrumentation employed. If, for instance, the requirement is to investigate trends in air pollutant concentrations with time, then the monitoring equipment must be capable of resolving differences in concentration at least as small as the expected increase or decrease arising from the trend in time. If, however, the study aims to quantify the spatial pollutant distribution in an urban environment, then an analyser capable of resolving to parts per trillion concentration levels would be no more useful than an analyser which resolves to parts per billion.

2.3 Choice of Representative Locations

If a study aims to assess short-term maximum concentration exposures of particular people, then the selected analysers will need to be situated in the most polluted locations where there is population exposure. Examples of this would be at the kerbside and in workplace environments. If, however, the study aims to assess the overall dosage of the majority of an urban population, then a background location, not directly influenced by specific sources or sinks, will be more representative of average population exposure.

3 Selection of Representative Monitoring Locations

After the objectives of the study and the site location criteria have been defined, the locations of individual sites can be considered.

We will take, as an example, the UK Department of the Environment's Rural Automatic Network used for monitoring atmospheric ozone, nitrogen oxides and sulfur dioxide. One of the objectives in operating this coordinated network of monitoring sites is that it enables comparisons to be made of data between monitoring sites. This comparison, however, is only valid if the sites are located in areas which are at suitable distances from pollution sources and sinks, or within similar areas of the environment. Pollution monitoring networks, therefore, should have a well defined set of site location criteria, to which all sites in the network adhere. Examples of site selection guidelines that have been adopted in the above Network are:

- The sampling height should be between 3 and 5 metres above ground.
- There should be unrestricted atmospheric flow towards the sampling head for at least 270 degrees in the horizontal plane.
- The sampling head should be at least 1 metre vertically from any other significant structures.
- The horizontal distance from the sample inlet to any solid object of significant dimensions should be at least five times the vertical distance between the inlet point and that object.
- The distance between the monitoring site and the nearest road should be at least 20 metres. The distance between the monitoring site and the nearest road carrying in excess of 10,000 vehicles per day should be at least 50 metres. The distance between the monitoring site and any road carrying in excess of 20,000 vehicles per day should be at least 100 metres.
- The site should be situated generally upwind of any significant combustion sources, at a distance of not less than 200 metres.
- The site should be at least 10 metres from the dripline of adjacent trees.

It may not be possible in practice to satisfy all the above site selection criteria at one location. For example, in rural locations the major constraining factors will be the availability of electrical power and telemetry links. In urban locations, however, the major constraints on site locations will be associated with avoiding specific pollutant sources and minimising the risk of vandalism.

It is important to ensure that the siting guidelines outlined above continue to be met at subsequent times, as well as ensuring that the potential site is satisfactory prior to installing the equipment. The results of not adhering to site selection criteria are likely to be that data gathered may not be representative and may therefore be challenged.

4 Selection of Monitoring Instrumentation

A wide range of different types of instrument is available for the measurement of ambient air pollution. The quality assurance and control system of any monitoring programme should include an assessment of the optimum most cost-effective instrumentation that will provide results to the required accuracy and with the maximum specified data capture rate. In principle, instruments for measuring air pollution vary, for example, from simple and inexpensive diffusive samplers through to expensive open-path monitors. Thus, the choice of the correct instrumentation is important if the effectiveness of the study and value for money are to be maximised. For example, when monitoring to establish compliance with the EC Directive on ambient NO_2, results are required with at least hourly resolution. Furthermore, this Directive specifies that ozone chemiluminescence should be used as the measurement method - thus restricting the choice of instrumentation. There are, however, a number of manufacturers of traditional 'continuous *in-situ*' analysers. Therefore, when considering which analyser to select, cost will clearly be a major consideration. However, consideration should also be given to the total cost of operating the instrument over its life. This will cover:

(a) **The level and costs of after-sales support:** the frequency of analyser service visits during the monitoring programme, and the tasks to be completed while carrying out these services, should be established.
(b) **Ease of operation.**
(c) **Reliability:** which may be assessed by contacting users of the instrument under consideration.
(d) **Results of type approval:** in certain countries, a procedure for 'type-testing' is in place, whereby manufacturers submit their analyser to an independent organisation for testing. These tests determine the instruments compliance with pre-set criteria covering, for example, for analyser noise, linearity, drift and response times.

Other factors which should be considered involve the selection of the associated infrastructural equipment. A summary of some of these items is given below:

4.1 Autocalibration Devices

Daily autocalibrations, or instrument checks, are a useful means of monitoring the data quality from analysers which operate unattended over extended time periods. Examples include permeation devices for calibrations of NO_x and SO_2 analysers and air scrubbers which are used to check instrument zero responses. In these cases when considering the type of autocalibration devices to acquire, parameters, for example, such as the time that the air scrubber and permeation tubes are specified to last before replacement becomes necessary should be

Table 1   *Autocalibration units used in the DoE Rural Monitoring Network*

	Ozone	*Sulfur dioxide*	*Nitrogen dioxide*
Zero	on-board charcoal scrubber	autocal unit: charcoal scrubber	autocal unit: 'purafill/charcoal' scrubber
Span	on-board ozone generator	autocal unit: permeation tube	autocal unit: permeation tube

considered. As a specific example, the type of autocalibration devices used in the UK DoE Automatic Rural Monitoring network are listed in Table 1.

It should be noted, however, that as the permeation tubes are not accurately or regularly weighed at sites, and the flow rates through the systems are not calibrated, these devices cannot be used to produce accurate quantitative data scaling factors. They are, however, useful in providing regular diagnostics on the operation of a continuous analyser. In the future, accurate calibration gas mixtures in cylinders, with their outputs controlled by solenoid valves, could be used as an alternative to permeation tubes.

4.2 Sample Manifold Systems

The sample manifold system is a particularly important component of the measurement system. The manifold system must be capable of allowing the ambient air to be transported to the inlet of the analyser without causing a significant proportion of the measurand to be lost by absorption. The detailed design of an appropriate sampling system is too complex to discuss here. However, two simple guidelines are that the sample residence time should be less than a few seconds, and that the pressure drop through the sampling system should be less than 60 Pa. In practice, to ensure an inert sampling system, inlet manifolds are generally constructed from glass, or from teflon or teflon-coated metal. Glass clearly has the advantage that it is possible to see clearly when the manifold is becoming contaminated. Its disadvantage is its fragility. A mechanical pump is generally used to provide the required throughput of atmospheric sample through the system. If possible, the output of this pump should be directed outside the equipment container so as to minimise the possibility of the flow through the manifold being affected by pressurisation inside the container arising, for example, from the use of an air conditioning unit.

4.3 Chart Recorders

Chart recorders are an important aid in assuring the quality of ambient air pollution data. This is because current data logging systems generally produce average results over a fifteen minute or an hourly average period. These averaging periods tend to mask short-term instrument response anomalies such as cyclic response, noisy output data, or intermittent signal drop-out. All of these potential problems, which if present could lead to data quality being compromised, are likely to go unnoticed if chart recorders or similar recording devices with high data capture rates are not used.

As an example problems occurred with particulate monitoring instruments (PM_{10}) within the DoE Enhanced Urban Network when it was first established. These were only diagnosed by evaluation of the backup chart recorders. These traces showed that the analysers had noisy signals, sometimes with noise in excess of the typical range of ambient concentrations, with a cyclical response of a similar magnitude to the typical ambient variations. This would have gone unnoticed from consideration of even 15 minute average data. As a result of inspection of the chart recorder record, the problem was identified and resolved.

A further benefit of chart recorders is that they make the calibration process more objective, by assisting the operator in determining when an analyser has stabilised on a calibration sample, without which there may be a degree of subjectivity as to when the reading is stable, thus potentially compromising the calibration.

5 Training and Auditing of Personnel

The most cost effective means of ensuring that high data capture is obtained from a network is to utilise personnel based within easy distance of the site to carry out calibrations and other routine activities. These local personnel, often referred to in the UK as Local Site Operators (LSOs), are also well placed to investigate certain suspected problems with the monitoring equipment. However, to ensure that the LSOs perform their duties in a rigorous manner throughout the network, they should undergo some form of appropriate training course. The training course should cover all aspects of their duties, with 'hands-on' training on the actual analysers they will be expected to work with. It is also advisable that they work to an appropriate operating procedure manual to ensure that a common reliable format is used in all their operations.

Training courses are held for new site operators in all DoE funded networks in the UK. It is part of the function of the network QA/QC unit to provide this training and to compile a suitable LSO procedure manual. In a similar manner, it is important that the organisation responsible for equipment servicing and repair should follow documented procedures, and complete pro-forma record sheets, to enable a retrospective evaluation of their actions.

In addition, in keeping with the requirements of a comprehensive quality system, such as that specified by the UK Accreditation Service (formerly NAMAS), all other aspects of the work should be documented. Therefore in addition to providing manuals covering the duties of the LSO, the QA/QC unit should compile manuals covering its own duties.

Having allowed some time for the LSOs to become well acquainted with their routine activities, a system for assessing and auditing their performance should also be carried out. Within the DoE Rural Monitoring Network this is done by an audit visit by the QA/QC Unit on an annual basis, at which the LSO is asked to perform all of their routine functions. The auditor will assess how competently these are being carried out, and an audit visit report will be written and provided to the ISO for guidance. In addition, as the QA/QC unit performs the audit visit, this provides a cost-effective opportunity to carry out additional analyser calibrations. Therefore, as part of the DoE Rural Network activities, analysers are tested against nationally-traceable standards at the same time as the audit visit as a further check to ensure that accurate data are being obtained. These audit visits also allow LSOs to ask questions or obtain clarification on matters concerning their work. This is particularly useful for new or substitute LSOs who have not yet been trained directly.

6 Calibration Methodologies

The calibration methodology provides the data required to convert the output voltages from the analysers into pollutant concentrations. The methodology to be used is generally specified by the QA/QC unit. It ultimately requires a reference to known concentration gas standards and calibration techniques, which have been validated at national or international level.

Calibration requirements differ between instrument type, and sometimes from site to site. The reason for frequent calibration is generally due to the degradation of the internal components of an analyser. For example, optical windows become clouded, pumps degrade and are not able to sustain the specified vacuum within a reaction chamber, and photomultiplier detectors alter their response with time. The above factors are examples where the response of a given analyser can degrade either rapidly or slowly with time.

These occurrences, however, are not intrinsically a problem as, providing that the calibration frequency is sufficient to fully quantify their effects, a high quality dataset will be produced.

The frequency of calibrations may vary significantly, depending on the type of monitoring programme being undertaken and the type of analysers in use. Typically, in an urban environment calibrations are carried out either weekly or fortnightly, whereas in rural locations the calibration frequency is longer, usually one month. This is because analysers are more likely to be adversely affected by exposure to urban rather than rural environments.

It is difficult however to produce repeatable test concentrations of ozone routinely in the field with which to check analyser response at weekly or monthly intervals. However, the standard measurement method for ozone, ultraviolet absorption, is for most commercial instruments intrinsically reliable. The measurement involves the determination of the intensity of ultraviolet radiation on a detector with and without ozone present in the sample cell - as the measurement depends on a ratio, most factors which would cause a drift in output are 'ratioed out'. Therefore, provided that certain components such as the ozone scrubbers are operating adequately, ozone analysers are

Table 2   *Calibration regime for the UK DoE Rural Monitoring Network*

	Ozone	*Oxides of Nitrogen*	*Sulfur dioxide*	*Particulates (PM_{10})*
daily	auto check, zero and span	auto check, zero and span	auto check, zero and span	none
monthly	manual activation of internal auto checks standards	manual zero and calibration with NO and NO_2	manual zero and calibration with SO_2 standards	none
6 monthly	photometric calibration, analyser noise, linearity, and response time tested.	calibration against nationally-traceable NO and NO_2 standards. Analyser noise and linearity converter efficiency, response time, range ratio tested	calibration against nationally-traceable SO_2 standards. Analyser noise and linearity, hydrocarbon interference, response time, range ratio tested	flow measurements; multipoint verification of correct calibration constant

stable - with typical variations of less than $\pm 1.5\%$ per six months. As such, their response may be adequately quantified by calibration against a 'transfer standard' ultraviolet photometer at six monthly intervals. In the case of the Rural Network, this ultraviolet photometer is itself verified regularly against similar standards held in the UK and in other national laboratories. A primary ozone photometer of the type developed by the National Institute of Standards and Technology, USA, is one of the primary standards held by the National Physical Laboratory to underpin UK ozone measurements. The calibration regime presently operated in the DoE Rural Monitoring Network is shown in Table 2.

These calibrations in the field also provide an opportunity to test the internal diagnostics of the analyser as well as carrying out infrastructural checks. The procedure for routine site visits and analyser calibrations is generally carried out as outlined below.

6.1 Routine Calibrations

Routine calibrations of continuous analysers are generally carried out or at two more points - including a zero and single span point. The application of this linear calibration assumes that the analyser output varies linearly with concentration at the input. (This assumption of linearity is normally verified every six months by the QA/QC unit). It is imperative that all analyser outputs are noted from the medium on which the data are stored. Thus, if data are stored on a data logger, analyser outputs must be read from the logger.

It is also important that as much of the sampling system as practical is tested as part of the calibration. Thus, zero and span gases should be introduced to the analyser through the normal sampling port. (It should be noted that an ideal calibration procedure would involve injecting zero and span gas through the inlet manifold at the required flow rate. This would then test both the analyser response and determine whether losses occur in the sampling system. However the production of reliable zero and span gases at the required flow rates is difficult to achieve in the field and this therefore precludes this method of calibration.) 'Zero air' is normally produced by chemical scrubbing of the pollutant of interest and any potential interferents. A calibration gas sample, normally comprising a compressed gas mixture of accurately known concentration and verified stability in a gas cylinder, is introduced to the analyser, and the analysers response is noted after a stabilisation period.

Normally, it is possible to introduce span gas with the analyser set to the same concentration range as that on which ambient measurements are carried out. On occasion, however, the range must be changed to a 'calibration range' to note the analyser response to the span gas. In this event, a second zero must be taken on this new 'calibration range' and a more complex procedure is then required to produce valid, accurate results.

It is also important to note that the analyser is not routinely adjusted to agree with the concentrations of the calibration standard. Instead, the calibration results are recorded and any drift in response is taken account of in the new zero and span factors. The reason for not adjusting the analyser response is that it would alter the analyser response history. When an analyser drifts over time this can be accounted for when the final data are produced. However, this process is made more difficult and subject to errors when regular adjustments to the analyser's response is made. Furthermore, analysers may take some hours to stabilise fully after such adjustments and the data during these stabilisation periods is generally of lower quality than would otherwise have been the case.

6.2 Post-calibration Checks

Following analyser calibration, checks are made to verify that the analysers are not suffering from significant leaks. Following this, the sample inlet filters are changed where these are present. This is one reason for the increased frequency of visits at urban locations - the inlet filters, which protect the analyser's internal components from ambient particulates become clogged because of heavier particulate loading. The final stage of the visit is designed to ensure that the system is left running in the proper state to collect ambient data. For example, checks are made on the manifold system to check that the blower is functioning, and that the analysers have been properly reconnected to the inlet manifold.

6.3 Non-routine Calibrations

The form of the calibration in the event of analyser malfunction should also be specified. Where it is suspected that an analyser is malfunctioning, and either the service/repair organisation or the LSO has been called out to investigate the problem, the instrument should be calibrated as found - i.e. before any action has been taken to remedy the fault. By so doing, data capture may be maximised as, provided the fault has been well documented and its effects quantified, it may be possible to recover data obtained before the repair accordingly. On the other hand, if this calibration is not carried out, the effects on the instrumentation will remain unknown, and further data are likely to be lost. In some circumstances, however, a pre-repair calibration will not be possible due to the analyser having suffered a major breakdown. Nevertheless, it may be possible to provide a meaningful calibration after the analyser has been repaired in some circumstances. For instance, an ozone analyser should have the same response characteristics independently of its main power supply. Thus if a power supply fails and has to be replaced a calibration following such a replacement would be worthwhile. In all circumstances, instrumentation should always be calibrated before it is used following repair or replacement.

7 Traceability of Measurements to National Standards

The most cost-effective and transparent method of ensuring that measurements from a particular survey are comparable with measurements carried out elsewhere is to ensure that the data are consistent with the standards maintained as part of the National Measurement System. To achieve this, the measurements must ultimately be traceable to National Standards, generally held in the UK at the National Physical Laboratory. Having established traceability to National Standards, which are regularly compared with those of other countries, the measurements are then comparable with all other traceable measurements in the UK and abroad.

NPL maintains primary gas standards covering a large number of species over a wide range of concentrations. These mixtures are prepared gravimetrically to the highest metrological standards, and are extensively intercompared with established primary standards. They are also rigorously analysed for impurities and regularly compared with similar standards held by other countries. Primary national standards held by NPL used in the field of air quality monitoring include:

Nitric oxide: in the range 200 to 500 ppb, prepared gravimetrically.

Nitrogen dioxide: at the sub-ppm level, prepared using a continuously-weighed permeation tube system and validated by other gravimetric techniques.

Sulfur dioxide: at the sub-ppm level, prepared using a continuously-weighed permeation tube system, and validated by other gravimetric techniques.

Ozone: NPL use a reference photometer as a primary ozone standard, obtained from NIST (US). This is routinely checked against a second instrument to verify stability.

Hydrocarbons: Standards containing 27 hydrocarbon species at accurately known concentrations (10-100 ppb), similar to those found in ambient air, prepared gravimetrically and validated by intercomparisons around Europe.

Standards for ozone and nitric oxide are inter-compared by means of gas phase titration, which utilises the gas phase reaction

$$O_3 + NO \rightarrow NO_2 + O_2$$

Thus, by reacting NO and O_3 with NO in excess, the amount of NO lost (to form NO_2) stoichiometrically equals the amount of O_3 added. Therefore, if an ozone analyser measures the ozone added, and an NO_x analyser measures the amount of NO lost as a result of this addition, and the change in NO equals the concentration of ozone, then the standards against which the analysers were calibrated may be related. This reaction has been carried out at NPL, with ozone and NO standards agreeing to within $\pm 1\%$ to 2%, thus providing a high degree of confidence in the validity of the ozone and NO standards.

Calibrated measurements of gas flow are also required to ensure the quality of ambient measurements. Flow measurements carried out at NPL are calibrated by reference to the gravimetric loss of gas from a cylinder. A cylinder of gas is allowed to flow through a mass flow controller (MFC) at a constant rate for a known time. The cylinder is weighed before and after the measurement enabling the flow to be determined from the mass loss of the cylinder. By carrying out a number of such measurements, at different flows, the MFC is calibrated over the required range. This gravimetrically-derived standard is then compared routinely with a volumetric calibrator which measured the time taken for a frictionless float to traverse an accurately known volume.

A supplementary method for ensuring the accuracy of standards that underpin the air quality monitoring programme is to intercompare these directly with standards held by other laboratories. In the case of the DoE UK Rural Monitoring Network, this has been carried out in a number of stages.

Within the UK, standards are checked every six months with those held by AEA Technology. In these instances, AEA make estimates of the concentration of the on-site cylinders of NO, NO_2 and SO_2, at a rural network site, as well as determining a network ozone analyser response in relation to the photometer held by AEA. Similar comparisons are made by NPL, the results of the two laboratories determinations being combined to form the overall intercomparison. The comparisons between NPL and AEA (currently the QA/QC unit for the DoE Automatic Urban Network) ensure that uniformity is maintained within prescribed limits between data sets from both the urban and rural networks - as well as establishing the extent to which AEA measurements are traceable to UK standards.

NPL have also compared standards with laboratories involved in the maintenance of environmental standards in other countries. The NPL ozone photometer has been compared with those of RIVM (the Netherlands) and the European Joint Research Centre (JRC) at Ispra. Comparisons carried out thus far have demonstrated the accuracy of the NPL system to within $\pm 1\%$. Nitrogen oxide standards have been compared with those at NMi and RIVM (The Netherlands), and the National Institute of Standards and Technology (USA), and sulfur dioxide standards with RIVM and NMi.

8 Ensuring Intercomparability of Measurements throughout a Network

As noted previously, air pollution monitoring networks allow direct comparison of data from different locations. A simple and effective way of ensuring that data from individual sites are directly comparable is to carry out a network intercomparison. Such exercises are regularly carried out in UK monitoring networks. The basic procedure is as follows:

8.1 Ozone

A standard photometer is transported to each monitoring site. Six concentrations covering the range 0 to 200 ppb are produced and introduced to the analyser. After sufficient stabilisation time has occurred the response of the analyser is noted from the datalogger voltage outputs. Then, as the data are taken from the on-site datalogger, the response of the analyser is evaluated as a linear or nonlinear regression between photometer concentrations in ppb and the analyser voltage outputs. By comparing each network analyser against a single transportable standard, this gives a valid indication of the comparability of data throughout the network.

8.2 Oxides of Nitrogen, Sulfur Dioxide and Carbon Monoxide

(i) A stable 'intercalibration standard', which has been validated against UK primary standards, is transported to each site and is injected into the appropriate analyser.

(ii) A 'site estimate' of the concentration of this standard is produced by scaling the analyser's output using the zero and span factors obtained from the most recent calibration. The scaling data are supplied from the network management unit. In this way the efficiency of the personnel carrying out the calibration, as well as the interpretation of the calibration utilised by the management unit, is included in the procedure which is used to derive the intercalibration results.

(iii) The site estimates from all the sites are then averaged, and the results from each site are compared with the mean. An 'outlier' is defined as a result with a deviation of greater than $\pm 10\%$, or two standard deviations, from the mean.

(iv) If outliers are identified the reason for the deviation is investigated rigorously, and the mean of site estimates are recalculated - either ignoring the outlier or incorporating the newly corrected result from the outlying site.

(v) The mean of site estimates is then compared with the primary determination of the concentrations of the intercalibration standard by expressing the results as a single ratio. This therefore gives an indication of the overall accuracy of the complete Network data set, in a manner which is traceable to National Standards.

These intercalibrations are generally carried out every six months in conjunction with analyser servicing schedules.

9 Performance Checks on Monitoring Instruments

As has been noted above, the performance of continuous analysers and other sampling instrumentation can change over time. For this reason, it is important that monitoring instrumentation is regularly checked. With continuous analysers, the regular calibrations act as the main check on analyser performance, but there are a number of parameters which - in addition to routine zero and span determination - must be checked, to allow the data to be considered as being of high quality. The parameters which are most likely to affect the quality of data from continuous analysers are outlined below.

(a) Linearity

This is a measure of the validity of the procedure used to scale the data linearity as described in Section 6 above. To determine analyser linearity, a series of concentrations are produced which cover the analyser range. A linear regression is then carried out, relating analyser output to the concentration being produced. The linearity error, defined as the standard error of the regression slope should be less than $\pm 2\%$ relative.

(b) Range Ratio

This is only measured in the case of having to calibrate analysers on a different range from that on which they are normally run. This test is performed by producing a concentration within the analyser running range, noting its output, and changing the analyser range to the 'calibration' range, again noting the output. The ratio of the two - appropriately zero corrected - outputs, should be within $\pm 1\%$ of the 'nominal' range ratio - given by the ratio of the two full-scale concentration values.

(c) Response Time

This defines how rapidly the analyser responds and stabilises to a step change (increase and decrease) in a concentration at its inlet. For hourly averaged data, analyser response times are generally well within the data averaging period. For certain studies which involve, for example, one minute averaging, analyser response times are likely to become significant.

(d) Noise Level

This is an indication of how reproducible, over short periods, the analyser is to a constant concentration at its inlet. For hourly averaged data, it is probable that any analyser noise will be averaged out satisfactorily. For shorter averaging times, however, an analyser showing output noise at a significant level is likely to increase the overall uncertainty of the measurement.

(e) Converter Efficiency

This is a particularly important parameter in the determination of NO_2 by the ozone chemiluminescence method. In a chemiluminescent NO_x analyser, NO_2 is converted to NO in a molybdenum converter, prior to measurement. Obviously, if the converter is inefficient, this will result in the erroneous measurement of NO_2. It is also important in the determination of methane/non-methane hydrocarbon analysers. These converters may be tested as follows:

A stable concentration of NO is produced and the analyser outputs, NO_x and NO, are noted after a suitable stabilisation period. Ozone is then added to the sample, converting some of the NO to NO_2, in a manner where the total NO_x in the sample remains constant. The NO_x and NO outputs are again recorded. The converter inefficiency is defined as the ratio of the change in scaled NO_x signal to the change in the scaled NO signal, usually expressed as a percentage. It is recommended that converter efficiency should be greater than 96%.

(f) Interference Tests

The most commonly used interference test checks the effect of hydrocarbons on the response of SO_2 analysers. The test checks the efficiency of the hydrocarbon removal facility since it is known that certain hydrocarbons affect the ultra-violet fluorescence measurement method.

(g) Tests of Particulate Analysers

Particulate analysers and their sampling systems generally make a measurement of mass collected on a filter, combining this with an assumed or measured flow to produce a particulate mass concentration expressed in microgrammes per cubic metre. It is important to verify both the mass measurement and the analyser flow-rate. Mass measurement may easily be verified by monitoring the instrument response to introduction of pre-weighed filters. Flow rates may be measured by transporting calibrated flow measurement systems to monitoring sites and comparing actual measured flows with those assumed by the sampling system. It is important that such flow monitors are accurate and preferably traceable to national standards. Other analyser response parameters, such as zero and span drift are generally quantified by consideration of long-term calibration data.

In the UK DoE Rural Monitoring Network, the above checks are regularly carried out at each monitoring site, in conjunction with routine analyser servicing as follows.

By employing a system such as this, the QA/QC unit of a network are able to quantify analyser performance before and after instrument service. The pre-service visit should be considered as providing an 'end of period' calibration whilst the post-service visit provides a 'beginning of period' calibration. As well as verifying the integrity of the monitoring instruments, it is essential that the sampling system is checked. This is done by

(i)   verifying that the analysers are connected to the sample manifold
(ii) measuring the flow-rate through the manifold
(iii) measuring the pressure drop inside the manifold
(iv) visually inspecting the manifold for signs of contamination or blockages which might affect the concentration of pollutant species.

As noted previously, an ideal situation would be one in which calibration or test gases could be passed down the manifold to quantify the effects of the manifold. Such systems are not in routine use in the UK although some research is in progress.

10 Production of a Final Data Set

The production of a finalised data set is, generally known as 'data ratification'. This is the process whereby 'raw' or 'provisionally validated' data are combined with all other relevant information to generate the final accurate dataset. This process reviews a multiplicity of inputs to provide a traceable and reliable dataset. This function is generally carried out by the network QA/QC unit. Factors considered during the ratification process are:

- Daily automatic zero and span results.
- Routine and non routine manual calibration data.
- Network intercalibration results and detailed instrument checks.
- Data regarding analyser service and breakdowns.
- Backup chart recorder traces.

All calibration data are plotted over a long time scale to allow analyser response trends to be assessed. Meteorological data and pollution data from other monitoring sites are also used where appropriate.

A suite of computer software is used to manipulate and display the large volumes of data generated by a particular network. Sophisticated statistical techniques are also used. However, as it is not possible to automate all functions, such as the interpretation of calibration data and the screening of data for subtle instrument faults, these generally are carried out by experienced scientists within the QA/QC Unit.

The overall procedure for ratifying air quality data, as exemplified by the work carried out for the UK DoE Rural Network, is outlined below.

Data are provided by the Central Management and Coordination Unit in three-month blocks. This includes daily automatic calibrations and also daily values for the calibrations used by CMCU in producing the initial data. These data are then formatted to conform with the NPL data ratification system.

The next stage is pre-ratification. This is the point where all relevant records, as well as the monthly datasets, are scrutinised and initial scalings, corrections and deletions are made. A computer program combines one month's raw data with the relevant calibration files - both manual calibrations and automatic analyser 'check' calibrations for a particular pollutant from a single site. Each of these monthly files is contained within a spreadsheet, which include the above data, and plots of both raw and 'scaled' data. Data are initially scaled using the calibrations as derived from the routine calibrations. At this stage the validated data - as produced by the management unit for rapid release to the public via teletext services etc - are discarded. Since the ratified data are derived from the raw data, this avoids potential problems with propagating errors in the validation process.

Data are manipulated as 15-minute averages, with their time stamp corresponding to the end time of the averaging period.

Each monthly spreadsheet is set up to allow zero and span values to scale each column of daily data. This is carried out by, initially, using the relevant manual calibration factor.

Single channel analysers (SO_2 and O_3 for example) are usually scaled linearly and a zero offset applied. In the case of NO_x analysers, however, each instrument channel should be scaled independently, with NO_2 being derived from their arithmetic difference.

Special procedures are employed to cope with, for instance, a change of analyser at a site, necessitating different span responses within one day.

As part of the spreadsheet ratification system, the data are assessed to ensure that

- all calibrations are consistent and have been applied to the raw data correctly.
- negative hourly average data in excess of pre-determined limits are not present.
- the analysers have not malfunctioned.

If data are nulled, re-scaled or re-zeroed for any reason this is documented. Any such changes will also initiate re-plotting of ambient and calibration data. Should a linear or nonlinear function be necessary to correct an analyser span coefficient, the interpolated span value is generally applied on a daily basis. If, however, the overall magnitude of the ramp is small, a linear interpretation will be applied on a monthly basis.

When the data have been inspected, taking account of the above items, the fifteen-minute averaged data are converted into hourly averages. This is also carried out under computer program control using validated software.

This program outputs the data from the spreadsheet in two formats:

- A monthly pollutant spreadsheet, containing raw and scaled fifteen minute averages, hourly averages, all the automatic and manual calibration data, data plots, and scaling formulae.
- Hourly averages data, with relevant site, pollutant and period identifiers. This may then be loaded into the NPL data archive used in reports.

Plots of the complete six-monthly data sets are collated and circulated amongst senior project team members with complementary expertise, who review the data. These data sets, including comments, are returned to the officers responsible for ratification who then re-examine the data in the light of the comments. The data are finally ratified at a meeting of senior project team members, who, after detailed discussion on each apparent problem, reach a consensus on the most appropriate action to take.

11 Measurement Uncertainty

The assignment of uncertainty to measured values is currently a topic of much international discussion. In the past, particularly in the area of analytical measurements of which air quality monitoring forms a part, uncertainties, or values for accuracy and precision, have tended to be ascribed on an *ad hoc* basis. There has been considerable effort devoted recently to standardising the calculation and expression of uncertainty so that the values for different measurements can be compared reliably. The basic guidelines were set out in the 'Guide to the Expression of Uncertainty in Measurement' published by ISO (the International Standards Organisation) in the name of seven key international bodies in 1993.

For convenience it is often sensible to express uncertainties for data in general (as opposed to single data points) as a combination of % (due to uncertainty in the scaling factor) and an absolute number of, say, ppb (due to uncertainty in the zero). However, it is not technically correct to distinguish these as 'accuracy' and 'precision' respectively. They both relate to accuracy (closeness to the true value), rather than precision (closeness of repeated measurements of the same quantity).

In the case of air quality measurements within the control of NPL, a rigorous evaluation of uncertainty following the recent guidelines is in the progress of being carried out. It is clear that a proper estimation of uncertainty depends on all aspects of the QA/QC system, from instrument reliability to calibration to ratification procedures, so that giving typical values for different pollutants is likely to be misleading.

12 Statistical Analysis of Trends in Air Quality Data

An important application for data acquired from automatic monitoring stations is the evaluation of trends in air quality. In order to be useful for this application, the data must have been the subject of a rigorous quality regime that has been maintained consistently over the time periods of interest, which may be as long as 10 or more years. Trends in the measured mean concentration may be as small as 0.1 ppb per year, which not only imposes requirements on the QA/QC of the data, but also on the statistical methods used to evaluate these trends.

A range of sophisticated statistical techniques have been developed at NPL to evaluate trends in air quality data. Particular issues that these techniques can overcome include:

- evaluating a trend which may be considerably smaller than the diurnal and seasonal variations in the data.
- calculating a statistical confidence interval for the result that does not require the assumption that the data are normally distributed.
- aggregating the results from many different sites to provide a valid estimate of a trend over the whole network.

Figure 1 shows the results of an analysis of the annual trend in mean ozone concentration at the 16 sites of the UK Rural Ozone Monitoring network. The figure shows the trend at each site evaluated using four different statistical 'models'. These rely upon fitting a model with 1, 4, 12 or 52 seasonal factors and then evaluating the confidence interval by bootstrap re-sampling. In some cases the results are significantly different from the results obtained from a straightforward 'linear least squares' fit to the data.

13 Summary

The procedures used in quality assuring and quality controlling ambient air pollution monitoring data are complex and cover all facets of the measurement system. These QA/QC procedures cover monitoring programme design, the selection of monitoring sites and instrumentation, the calibration and testing of the instrumentation, the training and auditing of operators, the requirement to ensure the accuracy and traceability of calibration standards, and the subsequent final data review and ratification. If all the elements noted above are taken into account, the data obtained from a monitoring exercise will be comprehensively quality assured and the results obtained should be valid and

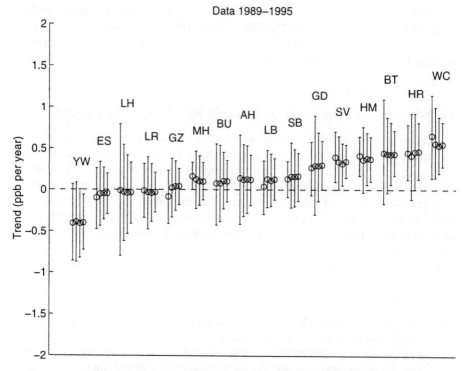

Figure 1 *Best estimates and 95% confidence intervals for trends in mean ozone concentration at the 16 sites of the UK Rural Ozone Monitoring Network. Each group relates (from left to right) to estimates based on a statistical model using yearly, quarterly, monthly and weekly factors together with a linear trend. The sites are ordered according to the resulting values.*

achieve the required accuracy. A brief discussion is presented on the accuracy of measurements achieveable in UK Networks. The results obtained in this manner may then be used to define long-term trends in atmospheric pollutant concentrations.

Subject Index

T
1 Month